ENTROPY IN CONTROL ENGINEERING

SERIES IN INTELLIGENT CONTROL AND INTELLIGENT AUTOMATION

Editor-in-Charge: Fei-Yue Wang
(*University of Arizona*)

Vol. 1: Reliable Plan Selection by Intelligent Machines
(*J E McInroy, J C Musto, and G N Saridis*)

Vol. 2: Design of Intelligent Control Systems Based on Hierachical Stochastic Automata (*P Lima and G N Saridis*)

Vol. 3: Intelligent Task Planning Using Fuzzy Petri Nets
(*T Cao and A C Sanderson*)

Vol. 6: Modeling, Simulation, and Control of Flexible Manufacturing Systems: A Petri Net Approach (*M Zhou and K Venkatesh*)

Vol. 7: Intelligent Control: Principles, Techniques, and Applications (*Z-X Cai*)

Vol. 10: Autonomous Rock Excavation: Intelligent Control Techniques and Experimentation (*X Shi, P J A Lever and F Y Wang*)

Vol. 11: Multisensor Fusion: A Minimal Representation Framework
(*R Joshi and A C Sanderson*)

Forthcoming volumes:

Vol. 4: Advanced Studies in Flexible Robotic Manipulators: Modeling, Design, Control, and Applications (*F Y Wang*)

Vol. 5: Computational Foundations for Intelligent Systems (*S J Yakowitz*)

Vol. 8: Advanced Topics in Computer Vision and Pattern Recognition
(*E Sung, D Mital, E K Teoh, H Wang, and Z Li*)

Vol. 9: Petri Nets for Supervisory Control of Discrete Event Systems:
A Structural Approach (*A Giua and F DiCesare*)

Series in
Intelligent Control and Intelligent Automation
Vol. 12

ENTROPY IN CONTROL ENGINEERING

George N Saridis
Rensselaer Polytechnic Institute, USA

Published by

World Scientific Publishing Co. Pte. Ltd.

P O Box 128, Farrer Road, Singapore 912805

USA office: Suite 1B, 1060 Main Street, River Edge, NJ 07661

UK office: 57 Shelton Street, Covent Garden, London WC2H 9HE

British Library Cataloguing-in-Publication Data
A catalogue record for this book is available from the British Library.

ENTROPY IN CONTROL ENGINEERING
Series in Intelligent Control and Intelligent Automation — Volume 12

ISBN 981-02-4551-3

Printed in Singapore by World Scientific Printers

To ***Dimitrakis***
a brave young man

TABLE OF CONTENTS

TABLE OF FIGURES

TABLES

PREFACE

The foundation of the present work was laid twenty five years ago in an office at the National Science Foundation in Washington DC. There, early every morning, with a cup of coffee, a serious discussion was taking place with Norm Caplan, about the future of Control Engineering. These discussions along with the interaction with Alex Levis, and Alex Meystel ended up with the conception of Entropy as a key measure for Control and Sensing theories which founded the theory of Hierarchically Intelligent Control and Robotics as its application.

This volume attempts to couple control engineering with the modern developments in science, through the concept of entropy. Such disciplines like Intelligent Machines, Economics, Manufacturing, Environmental Systems, Waste etc. can be favorably affected and their performance can be improved or actually their catastrophic effects minimized. Entropy is used as the unifying measure of the various, seemingly disjoint, disciplines to represent the cost of producing work that improves the standards of living, both in engineering and in science. Modeling is done through probabilistic methods, thus establishing the irreversibility of the processes involved. This is in accordance with the modern point of view of science. In addition the behavior of control for an arbitrary but fixed controller away from the optimal (equilibrium) was obtained, the analytic expression of which should lead to chaotic solutions. The control activity is explained herein, based on the principle that *control is making a system do what we want it to do*. This helps to relate control theory with the sciences.

This book is organized in the following way.

Chapter 1, introduces the concepts of entropy, chaos, and control engineering, to familiarize the reader with the key notions of the book. Chapter 2, discusses a new derivation of the theory of optimal estimation and control by associating the cost of performance with entropy. This new theory produced two results: first the derivation of the Generalized Hamilton-Jacobi-Bellman equation which describes the behavior of control systems for a fixed but arbitrary controller, away from the optimal (equilibrium) solution and second the analytic verification of Fel'dbaum's claim of Dual Control that the optimal closed-loop solution is obtained by the separation theorem *minus the active transmission of information.* Chapter 3, reviews the theory of Hierarchically Intelligent Control Systems. This work presented the motivation for using entropy as a measure of the cost function of such disjoint disciplines as control and sensing. The entropy formulation is presented for the three levels of the hierarchy: *Organization, Coordination, and Execution.* Chapter 4, defines *system reliability* as entropy. Using tight measures one may come up with simpler expressions which actually separate the cost of reliability from the cost of the experiment. Chapter 5, introduces entropy to a modern fully automated factory. It also demonstrates the application of Hierarchically Intelligent Control to production scheduling. Application of entropy derived control to environmental systems is the subject of Chapter 6. Here the principle of minimizing the entropy generated by human interaction with the environment is applied and demonstrates a way of the waste produced by generating work for the improvement of the quality of living. The systems under consideration are only four, Ecological, Biochemical, Pollution, and Econometric, but the concept can be extended to other non-engineering systems. Chapter 7, presents the application of Hierarchically Intelligent Control to a hardware oriented system; the unmanned construction of trusses for the Space Station. This project was funded by NASA, as a Center of Excellence and was performed at the Center of Intelligent Control Systems for Space Exploration (CIRSSE) at the Rensselaer Polytechnic Institute between 1987 and 1992. Finally, Chapter 8, gives some summarizing opinions of the whole project of coupling Control Engineering with the concepts of Entropy, and Chaos.

The author wants to gratefully acknowledge the contribution of all my students that have participated in the development of this work. I would also like to thank Norm Caplan, Alex Levis, Alex Meystel, and most of all my wife Youla for their encouragement, and constructive remarks during lengthy discussions.

George N. Saridis PhD
Athens Greece 1999

CHAPTER 1

ENTROPY, CONTROL, CHAOS

1.1. INTRODUCTION:.

From their primitive years, humans have tried to understand and formalize the world around them, through the sensors that nature provided them. To do that they have used various models to represent "approximations" of the functions of the world. They separated those functions into two categories governed by:

1.The Physical Laws of Nature describing deterministic physical phenomena, and

2. The Behavioral Laws describing nondeterministic organic, environmental and societal phenomena.

For the first class mathematics proved to be a concise methodology to approximately describe the time-reversible results of physical experiments, while logical (Aristotelian) analysis and statistical exhaustive search, were the methodologies used to classify and study the evolutionary behavior of living organisms, environmental and ecological changes and societal phenomena that demonstrated time-irreversibility (Brooks and Wiley 1988, Prigogine 1989).

In the recent years, with the progress of the mathematical science and the development of digital computers, probabilistic and stochastic methods and analytic logic are replacing statistical aggregation and classical logical analysis in the realm of mathematics to describe the world's phenomena. Linear analytic models were assumed to be sufficiently accurate to represent useful models of this world, as viewed by human senses. "Reductionism" that has been a powerful tool to analyze and predict physical phenomena, was promptly extended to cover natural phenomena for description and prediction of their behavior.

However, there were cracks in this wonderful and supposedly airtight, reasoning system. Physical discrepancies and analytic paradoxes marred the perfect models that the world thought infallible. One of the major difficulties encountered was the gap between Newtonian mechanics and thermodynamics. Scientists discovered that heat was produced by the collision of millions of particles in a perfect gas, generating irreversibly entropy, a lower level of energy. However, Poincaré showed, that it is practically impossible to study the motion of more than three bodies and thus understand the process. Boltzmann(1872) bridged this gap by introducing statistical methods to describe kinetic phenomena and equate their average kinetic energy with entropy. This pioneer work showed a way to model uncertain and complex physical phenomena in continuous time and connected them to irreversible evolutionary models described by Darwin (Brooks and Wiley 1988).

Shannon (1963) followed with his celebrated **information theory**.

Saridis (1995), in the meanwhile using Jaynes' principle (1957), formulated the engineering design problem as a problem with uncertainty, since the designer does not know a priori the outcome of his design. Entropy was used as the measure of the energy expressing the cost of the irreversible associated process. Considering control as the work and entropy as its cost, the optimal control problem was recast as an entropy minimization problem and the known expressions were reproduced. The cost of the reliability of the design was also expressed as entropy, and was considered as a natural extension of the proposed theoretical development.

In addition Saridis (1998), working on the problem of reducing analytically the increase of entropy generated by human intervention in ecological systems, added an extra term to the equation of ecosystems and used entropy formulation of optimal control to minimize its effect.

Major problems regarding the completeness, consistency and decidability of a statement in a discrete event space, arose with Gödel's theorem of undecidable statements that limited the use of digital computers for the analytic solution of complex problems. Such problems existed with the Diophantine equations and other paradoxes but they were swept under the rug, so that they would not challenge the power of the computer. Such problems were remedied by introducing new quasi-statistical engines like artificial intelligence, fuzzy set theory and other such techniques. All those problems were blamed to the inadequacy of linear models, and the complexity of systems operating away from the equilibrium point like most of the biological, environmental and societal systems do. Thus, the theory of Chaos was introduced by Prigogine (1996) to study and analyze such cases. The benefit of these discoveries was that complexity and undecidability brought all these problems together and global formulation of their solution was sought. Uncertainty, which is indirectly associated with time irreversibility, was the common element representing the doubt of the outcome of such systems, and stochastic approaches were introduced which have entropy as a common measure.

Thus, the uncertainty of modeling of complex systems is the reason of introducing entropy, in Shannon's sense, as a measure of quality of large complex continuous or discrete event systems. In essence, since entropy is energy irreversibly accumulated when work is performed, and originally introduced in thermodynamics, it is generalized to any kind of dynamic system appearing in nature. Therefore, entropy measures the waste produced when work is done for the improvement of the quality of human life, the struggle of the species in an ecosystem, the biological reactions of a living organism, even the politics of in a societal system. Entropy assumes a stochastic model with uncertain outcome, which is suitable to describe the new complex model of the world.

The question now is if and how this model and therefore the underlining systems, can be improved, by reducing the waste of energy represented by entropy, using analytic methods. Concepts from control theory, with the appropriate reformulation, is herein proposed to solve this problem.

1.2. GLOBAL ENTROPY

When we consume energy in order to accomplish some work in our environment we simultaneously generate a low quality residual energy, the cost of which irreversibly reduces the quality of the environment and leads to a chaotic situation. An infinite number of paradigms exist in our environment, starting with the pollution of the air, the water resources, traffic congestion, financial disasters, unemployment with the resulting crime, and in general the decay of the life-sustaining resources of mankind as mentioned by Rifkin (1989).

This low quality energy was discovered by the physicist Clausius which appeared in the second law of thermodynamics and was named Entropy. According to this law the production of work is followed by the production of residual energy that irreversibly increases the total level of the lower level energy and would potentially lead to thermal death.

A different interpretation of entropy was given by Claude Shannon (1963), as a measure of uncertainty in information theory related to communication systems. This interpretation was used by Saridis (1995), to introduce a theory which presents Automatic Control as a generalization of the theory of entropy, based on the designer's uncertainty to obtain the optimal solution. This concept is hereby extended to cover subjects related to the environment, finances, pollution and other problems that puzzle our present society.

1.2.1 REVIEW OF ENTROPY CONCEPTS

As previously mentioned, entropy is a form of low quality energy, first encountered in Thermodynamics. It represents an undesirable form of energy that is accumulated when any type of work is generated. Recently it served as a model of different types of energy based resources, like transmission of information, biological growth, environmental waste, etc. Entropy was currently introduced, as a unifying measure of performance of the different levels of an Intelligent Machine by Saridis (1985). Such a machine is aimed at the creation of **modern intelligent robots which may perform human tasks with minimum interaction with a human operator**. Since the activities of such a machine are energy related, entropy may easily serve as the measure of the cost of performing various tasks as Intelligent Control, Image Processing, Task Planning and Organization, and System Communication among diversified disciplines with different performance criteria. The model to be used is borrowed from Information Theory, where the uncertainty of design is measured by a probability density function over the appropriate space, generated by Jaynes' Maximum Entropy Principle.

Other applications of the Entropy concept are for defining Reliability measures for design purposes and obtaining measures of complexity of the performance of a system, useful in the development of the theory of Intelligent Machines.

Entropy is a convenient global measure of performance because of its wide applicability to a large variety of systems of diverse disciplines including waste processing, environmental, socio-economic, biological and other. Thus, by serving as a common measure, it may expand system integration by incorporating say societal, economic or even environmental systems to engineering processes.

In this Chapter, Entropy concepts will be mainly used for the reformulation of the theory of stochastic optimal control and its approximation theory, state estimation, and parameter identification used in Adaptive Control systems. It will prove Fel'dbaum's claim that the stochastic Open-Loop-Feedback-Optimal and Adaptive Control, based on the **Certainty Equivalence Principle** (Kumar & Varaiya 1986), (Morse 1990, and 1992), are not optimal. It will also provide a measure of goodness of their approximation to the optimal in the form of an upper bound of the Entropy missed by the approximation.

An application to the design of Intelligent Machines using performance measures based on Reliability and Complexity will demonstrate the power of the approach.

1.2.2 ENTROPY AND THERMODYNAMICS

The concept of Entropy was introduced in Thermodynamics by Clausius in 1867, as the low quality energy resulting from the second law of Thermodynamics. This is the kind of energy which is generated as the result of any thermal activity, at the lower thermal level, and is not utilized by the process.

It was in 1872, though, that Boltzmann used this concept to create his **theory of statistical thermodynamics**, thus expressing the uncertainty of the state of the molecules of a perfect gas. The idea was created by the inability of the dynamic theory to account for all the collisions of the molecules, which generate the thermal energy. Boltzmann stated that the entropy of a perfect gas, changing states isothermally, at temperature T is given by;

$$S = - k \int_{X} (\psi\text{-}H)/kT \exp\{(\psi\text{-}H)/kT\}\, dx \qquad (1.1)$$

where ψ is the Gibbs energy, $\psi = -kT$. In $\exp\{-H/kT\}$, H is the total energy of the system, and k is Boltzmann's universal constant. Due to the size of the problem and the uncertainties involved in describing its dynamic behavior, a probabilistic model was assumed where the Entropy is a measure of the molecular distribution. If p(x) is defined as the probability of a molecule being in state x, thus assuming that,

$$p(x) = \exp\{(\psi\text{-}H)/kT\} \qquad (1.2)$$

where p(x) must satisfy the "incompressibility" property over time, of the probabilities, in the state space X, e.g.;

$$dp/dt = 0 \tag{1.3}$$

The incompressibility property is a differential constraint when the states are defined in a continuum, which in the case of perfect gases yields the Liouville equation. Substituting eq.(1.2) into eq.(1.1) the entropy of the system takes the form,

$$S = -k \int_X p(x) \ln p(x)\, dx \tag{1.4}$$

The above equation defines Entropy as a measure of the uncertainty about the state of the system, expressed by the probability density exponential function of the associated energy.

Actually, the problem of describing the entropy of an isothermal process should be derived from the Dynamical Theory of Thermodynamics, considering heat as the result of the kinetic and potential energies of molecular motion. It is the analogy of the two formulations that led into the study of the equivalence of entropy with the performance measure of a control system. If the Dynamical Theory of Thermodynamics is applied on the aggregate of the molecules of a perfect gas, an Average Lagrangian I, should be defined to describe the average performance over time of the state x of the gas,

$$I = \int_{t0}^{tf} L(x,t)\, dt \tag{1.5}$$

where the Lagrangian L(x,t) = (Kinetic energy) - (Potential energy). The Average Lagrangian when minimized, satisfies the **Second Law of Thermodynamics**. Since the formulations eqs.(1.1) and (1.5) are equivalent, the following relation should be true;

$$S = I/T \tag{1.6}$$

where T is the constant temperature of the isothermal process of a perfect gas (Lindsay and Margenau, 1957). This relation will be the key in order to express the performance measure of the control problem as Entropy.

1.2.3 ENTROPY AND INFORMATION THEORY

In the 1940's Shannon (1963), using Boltzmann's idea, e.g., eq. (1.4), defined Entropy (negative) as a measure of the uncertainty of the transmission of information, in his celebrated work on **Information Theory**:

$$H = -\int_\Omega p(s) \ln p(s)\, ds \tag{1.7}$$

where p(s) is a Gaussian density function over the space Ω of the information signals transmitted. The similarity of the two formulations is obvious, where the uncertainty about the state of the system is expressed by an exponential density function of the energy involved.

Shannon's theory was generalized for dynamic systems by Ashby (1975), Boettcher and Levis (1983), and Conant (1976) who also introduced various laws which cover information systems, like the Partition Law of Information rates.

1.2.4 ε-ENTROPY

It implies that an increase in knowledge about a system, decreases the amount of ε-entropy which measures the uncertainty (complexity) involved with the system.

$$\varepsilon - H = \ln(n_e) \tag{1.8}$$

where n_e is the minimum number of coverings of a set e. Therefore ε-entropy is a measure of complexity of the system involved.

1.2.5 JAYNES' PRINCIPLE OF MAXIMUM ENTROPY.

In an attempt to generalize the principle, used by Boltzmann and Shannon, to describe the uncertainty of the performance of a system under a certain operating condition, Jaynes (1957) formulated his **Maximum Entropy Principle**, to apply it in Theoretical Mechanics. In summary it claims that

> **The uncertainty of an unspecified relation of the function of a system is expressed by an exponential density function of a known energy relation associated with the system.**

A modified version of the Principle, as it applies to the Control problem, will be derived in the sequel, using Calculus of Variations (Saridis 1988). The proposed derivation represents a new formulation of the control problem, either for deterministic or stochastic systems and for optimal or non-optimal solutions.

1.2.6 THE PRINCIPLE OF INCREASING PRECISION DECREASING INTELLIGENCE

In most organization systems, the control intelligence is hierarchically distributed from the highest level which represents the manager to the lowest level which represents the worker. On the other hand, the precision or skill of execution is distributed in an inverse manner from the bottom to the top as required for the most efficient performance of such complex systems. This was analytically formulated as the **Principle of Precision with**

Decreasing Intelligence (IPDI), by Saridis (1988). The formulation and proof of the principle is based on the concept of Entropy, and is discussed in Chapter 3.

1.2.7. ENTROPY AND THE ENVIRONMENT

Since the latest major improvements in the average quality of life, major increases have occurred in the production of waste, traffic congestion, biological pollution and in general environmental decay, which can be interpreted as the increase of the Global Entropy of our planet, an energy that tends to deteriorate the quality of our modern society. According to the second axiom of thermodynamics this is an irreversible phenomenon, and nothing can be done to eliminate it.

The intention of this work is to introduce optimal control, developed for systems engineering, to environmental systems to effectively restrain the growth of the Global Entropy. Since the paper is addressed to the nonspecialized reader, an attempt will be made to introduce the concepts of systems, automatic control, optimal control and global entropy for information purposes. Then a formal presentation will be made, of the proposed theory developed from an entropy point of view which will relate the optimal control theory to the Global Entropy, and thus present a method to minimize its effect to our society. This theory has in addition to the practical applications, a philosophical foundation that has implications to the quality of life and the future of our planet.

1.3 UNCERTAINTY AND THE CONTROL PROBLEM

The theory of Newtonian mechanics, which is the basis for the classical control theory, assumes, for fixed initial conditions, well defined deterministic motions which are reversible in time (Coveney, Highfield 1990). Thus any uncertainty appearing in the system, would eventually be reduced.

On the other hand, uncertainty has been associated with insufficient knowledge and lack of information in Thermodynamics and Information Theory. The models used were probabilistic and irreversible in time and thus not deterministically reducible.

Therefore, in modern physics, one may distinguish two categories of uncertainties:

- **Subjective (reducible) uncertainties**
- **Objective (irreducible) uncertainties**

The first category refers to the temporary lack of knowledge of the experimenter, while the latter is associated with phenomena from which no further improvement of knowledge is attainable.

Now, the control design problem will be associated to the subjective uncertainties by the following argument. The designer, is faced with the uncertainty of optimal design before tackling the problem, because he does not know yet the solution. He may assign an appropriate probability density function over the space of **admissible controls**, and reduce the design problem to the problem of finding the point (control law), that has **maximum probability of attaining the optimal value**. This approach will be pursued in the sequel with **Entropy**, the measure of the probability of uncertainty of the design.

In the literature, other attempts made to formulate the estimation problem (Kalata, Premier 1974), and the feedback control problem (Weidemann 1969), using different entropy approaches, found little success.

1.4. THE HUMAN INTERACTION

Ecological systems are analytically modeled as interactions among various species alone. The novelty of this approach is the introduction of a term which involves the human interaction u(t) with the system. For simplicity we assume that this term is linear.

Thus a **typical ecosystem** (Singh 1987) in an uncertain environment may be closely approximated by:

$$dx/dt = f(x,t) + G(t)w(t) + B(t)u(t); \qquad x(0)=x_0, \qquad (1.9)$$
$$z(t) = h(x,t) + v(t);$$

where x(t) is the state of the system at time t, z(t) is the current m-dimensional measurement vector, u(t) is the r-dimensional human interaction vector, f(x,t), h(x,t) are twice differentiable nonlinear functions of their arguments, B(t) and G(t) are nxr matrices of known coefficients respectively. The stochastic variables x(0) is the n-dimensional Gaussian initial state, w(t) is the r-dimensional process noise, v(t) is the m-dimensional measurement noise, defined for any $t\in[0,\infty]$, with properties:

$$E\{x(0)\} = x_0, \quad E\{(x(0) - x_0)(x(0) - x_0)^T\} = P_0, \qquad (1.10)$$
$$E\{w(t)\} = 0, \quad E\{w(t)w(\tau)^T\} = Q(t)\delta(t-\tau),$$
$$E\{v(t)\} = 0, \quad E\{v(t)v(\tau)^T\} = R(t)\delta(t-\tau).$$
$$E\{x(0)w(t)^T\} = E\{x(0)v(t)^T\} = E\{w(t)v(\tau)^T\} = 0.$$

We may want to use **Automatic Control theory** to control the human factor u, such that the system would reduce its bad influence.

1.5. AUTOMATIC CONTROL SYSTEMS

The methodology of Automatic Control Systems, briefly introduced here, has been applied to engineering systems for many centuries. The Theory Automatic Control is based on the concept of feedback (Brogan 1974). From the Watt regulator to the electronic cathode coupled amplifier, control system applications have been greatly effective. However, they became popular since the Second World War when they were used to control rockets. Their success was based on the principle of Feedback, which stabilized the performance of the system. In order to make things clear to all the readers, it is appropriate to clarify certain concepts of Modern Control Theory. This is done by the following definitions:

Definition 1.(System)

A system is every natural or other phenomenon that can be described by an analytic model, for the study, analysis, prediction and control of its behavior.

Definition 2.(Automatic Control)

Automatic Control is making a system do what you want it to do.

Definition 3.(Feedback)

In Automatic Control Systems a system is driven by the difference between its desired and actual output.

The above definitions are easily understood from the following analytic expressions to give a description of Automatic Control. They also represent a philosophical system that finds applications in every day life. Besides the modern electronic, electrical, mechanical or other engineering systems they may be generalized to include ecological, environmental, economic, administrative and other systems of general interest, by properly selecting the parameters of the analytic models of the respective system. Solution may be sought through standard control methods.

In the 1940's ,during the beginning of their applications, feedback systems, were influenced by communication technology, and were modeled in the frequency domain. Later, when state variables were introduced, they brought back the time domain models, which have the capacity of representing nonlinear, stochastic, digital, systems with discontinuities. Finally, optimal control was introduced in the 1960's as a mathematical tool to design automatic control systems.

A **typical system** in an uncertain environment may be closely approximated by:

$$dx/dt = f(x,u,t) + G(t)w(t); \qquad x(0)=x_0, \tag{1.11}$$
$$z(t) = h(x,u,t) + v(t);$$

where x(t) is the state of the system at time t, z(t) is the current m-dimensional measurement vector, u(t) is the r-dimensional control vector, f(x,u,t), h(x,u,t) are twice differentiable nonlinear functions of their arguments, and G(t) is an nxr matrix of known coefficients respectively. The stochastic variables x(0) is the n-dimensional Gaussian initial state, w(t) is the r-dimensional process noise, v(t) is the m-dimensional measurement noise, defined for any $t \in [0,\infty]$, with properties:

$$E\{x(0)\} = x_0, \quad E\{(x(0) - x_0)(x(0) - x_0)^T\} = P_0, \tag{1.12}$$
$$E\{w(t)\} = 0, \quad E\{w(t)w(\tau)^T\} = Q(t)\delta(t-\tau),$$
$$E\{v(t)\} = 0, \quad E\{v(t)v(\tau)^T\} = R(t)\delta(t-\tau).$$
$$E\{x(0)w(t)^T\} = E\{x(0)v(t)^T\} = E\{w(t)v(\tau)^T\} = 0.$$

Automatic Control implies the finding of a control u such that the system would behave according to our instructions. It may be open-loop where u(t) is preprogrammed and fed as an input, or feedback where the output is compared to a desired behavior and the difference is used to drive the system.

Optimal Control may be formulated as follows. A cost functional, containing all the predefined desired specifications about the system's performance is also defined:

$$V(u) = E\{J(u)\} = E\{\varphi(x(T),T) + \int_0^T L(x,u,\tau)d\tau\} \tag{1.13}$$

The solution of the optimal control problem is obtained by finding a control u^* that:

$$V(u^*) = V^* = \text{Min}_u\, I(u) \tag{1.14}$$

Optimal control represents philosophically the attempt of human nature to obtain the best results of meeting the given specifications. It may serve best when one attempts to reduce the Entropy of environmental decay.

1.6. ENTROPY FORMULATION OF CONTROL

It is desired to establish entropy measures, equivalent to the performance criteria of the optimal control problem, while providing a physical meaning to the latter. This is done by expressing the problem of control system design probabilistically and assigning a distribution function representing the uncertainty of selection of the optimal solution over the space of admissible controls. By selecting the worst case distribution, satisfying **Jaynes' Maximum Entropy Principle**, the performance criterion of the control is associated with the entropy of selecting a certain control (Jaynes 1957, Saridis 1985).

Then minimization of the differential entropy, which is equivalent to the average performance of the system, yields the formulation of optimal control. Adaptive control and stochastic optimal control are obtained as special cases of the optimal formulation, with the differential entropy of active transmission of information, claimed by Fel'dbaum (1965), as their difference. Upper bounds of the latter may yield measures of goodness of the various stochastic and adaptive control algorithms.

The Entropy formulation of Optimal Control, based on the modified Jaynes' Principle of Maximum Entropy (1957), was developed by Saridis (1995). Using the state equations and cost described by (1.11), (1.12) and (1.13) define the differential entropy, for some u(x,t),

$$H(x_0,u(x,t),p(u)) = H(u) = -\int_{\Omega x0}\int_{\Omega x} p(x_0,u)\ln p(x_0,u)\, du dx_0 \tag{1.15}$$

where $x_0 \in \Omega_{x0}$, $x \in \Omega_x$ the spaces of initial conditions and states respectively, and $p(x_0,u)=p(u)$ the probability density of selecting u. One may select the density function p(u) to maximize the differential entropy according to Jaynes' Maximum Entropy Principle, subject to $E\{V(x_0,u,t)\}=K$, for some u(x,t). This represents a problem more general than the optimal where K is a fixed but unknown constant, depending on the selection of u(x,t). The unconstrained expression of the differential entropy is;

$$I = \beta H(u) - \gamma[E\{V\} - K] - \alpha\left[\int_{\Omega x} p(u)dx - 1\right]$$
$$= -\int_{\Omega x}[\beta p(u)\ln p(u) + \gamma p(u)V]dx - \alpha\left[\int_{\Omega x} p(u)dx - 1\right]$$

Using the Lemma of Calculus of Variations, maximization of I with respect to p(u), e.g.

$$\partial H/\partial p = 0,$$

yields,

$$\partial/\partial p[-\beta p(u)\ln p(u) - \gamma p(u)V - \alpha p(u)] = 0; \quad \partial^2 H/\partial p^2 < 0$$

or

$$-\beta\ln p(u) - \beta - \gamma V - \alpha = 0; \qquad -\beta/p(u) < 0$$

and the worst case density is,

$$p(u) = e^{-\lambda-\mu V(u(x,t),x0,t0)}$$
$$e^{\lambda} = \int_{\Omega x} e^{-\mu V(u(x,t),x0,t0)}\, dx$$

with $\mu = \gamma/\beta$, $\lambda = (\alpha+1)/\beta$, and

$$E\{V(u(x,t),x_0,t_0)\} = K = \partial[\ln\int_{\Omega x} e^{-\mu V}\, dx]/\partial\mu \tag{1.16}$$

Now let us recast the optimal feedback control problem as the "optimal design" problem under the worst uncertainty of selecting the best control, from the set of admissible feedback controls, $u(x,t) \in \Omega_u XT \subset \Omega_x XT$. It is assumed that the set of admissible controls is covered by the probability density function p(u), expressing the uncertainty of selection of the optimal function. The worst case (maximum) differential entropy corresponding to p(u), is given by,

$$H(u) = \lambda + \mu E\{V(u(x,t),x_0,t_0)\} \tag{1.17}$$

The corresponding minimum value with respect to u(x,t), represents the optimal design, equivalent to the original minimization (eq. 1.14). This condition implies,

$$dH/du = \partial H/\partial u + \partial H/\partial p \; \partial p/\partial u = 0, \tag{1.18}$$

However, Jaynes' Principle worst entropy (1.17), which may replace the Performance Criterion, was based upon the condition,

$$\partial H/\partial p = 0.$$

which with (1.16), yields,

$$dH/du = \partial H/\partial u = 0 \tag{1.19}$$

This represents the solution of the optimal control problem in its entropy formulation.

Using the probability density function of eq. (1.16) for an arbitrary but fixed control u(x,t), and the incompressibility in time of the probability density condition, one may obtain the formulation of the control problem away from the equilibrium. The condition of incompressibility of the probability density:

$$dp/dt = 0 \tag{1.20}$$

yields an equation, equivalent to the Liouville equation (Prigogine 1980), which for the selected p(u(x,t)) is named the **Generalized Hamilton-Jacobi** equation

$$\partial V/\partial t + \partial V^T/\partial x \cdot f(x,u,t) + L(x,u,t) + \tfrac{1}{2}\mathcal{L}\{Tr(\partial V^2/\partial x^2 Qgg^T)\} = 0$$

$$V(T) = 0. \tag{1.21}$$

where $\mathcal{L}$ is the differential generator (Saridis and Lee 1979).

1.7 CONCLUSIONS

The evolution of modern science has introduced novel ideas to the interpretation of physical phenomena and integrated various disciplines like mechanics, thermodynamics, economics, biology, genetics, ecology, etc., that were previously considered as disjoint (Coveney and Highfield 1990). The result was to attempt to interpret global phenomena, in a unified way, using probabilistic descriptions with a common measure the underlying entropy, which permits their chaotic description in an analytic way away from their equilibrium (Prigogine 1996). This entropy measure was used by Saridis (1985) to represent the cost of performance of a system that does some work, in particular a control function. This approach casts the control problem as an uncertain activity which is in agreement with the methodologies of other disciplines studied within the theory of Chaos.

Time irreversibility and performance uncertainty enters the naturally the control activities.

Using this philosophy, optimal control can be introduced as the way of reducing the effect of undesirable human intervention in other disciplines of science. The resulting methodology is explored in this volume. It should be noticed that this presentation covers only a small area of the potential application of entropy to scientific thinking and further research should prove conclusive to the understanding of physical phenomena.

1.8 REFERENCES

Ashby W. R., (1975), *An Introduction to Cybernetics*,J. Wiley & Sons, Science Edition, New York.

Boettcher K. L., Levis A. H., (1983),"Modeling the Interacting Decision-Maker with Bounded Rationality", *IEEE Transactions on System Man and Cybernetics*, Vol. SMC-12, No. 3.

Boltzmann, L (1872) "Further Studies on Thermal Equilibrium between Gas Molecules" *Wien Ber.,* **66**, p. 275.

Brogan, W. L.,(1974) ***Modern Control Theory*** Quantum Publishers New York N.Y.

Brooks, D. R. and Wiley, E. O. (1988) ***Evolution as Entropy*** University of Chicago Press, Chicago Il.

Conant, R. C., (1976), "Laws of Information which Govern Systems", *IEEE Transactions on System Man and Cybernetics*, Vol. SMC-6, No. 4, pp. 240-255.

Coveney P. And Highfield R., (1990), ***The Arrow of Time*** , Fawcett Colombine, New York N.Y.

Feld'baum, A.A. (1965), ***Optimal Control Systems,*** Academic Press, New York.

Jaynes, E. T,(1957) "Information Theory and Statistical Mechanics", *Phys. Review*, **Vol.** 4, p.106.

Kalata, P. Premier, R., (1974), "On Minimal Error Entropy Stochastic Approximation" *International Journal of System Science*, Vol. 5, No. 9,
pp. 985-986.

Kolmogorov, A. N. (1956), "On some asymptotic characteristics of completely bounded metric systems" *Dokl. Akad. Nauk SSSR*, Vol. 108, no. 3, pp. 385-389.

Kumar, P.R., Varaiya, P., (1986), ***Stochastic Systems; Estimation, Identification and Adaptive Control***, Prentice Hall, Englewood Cliffs, NJ.

Lindsay, R.B., Margenau, (1957), ***Foundations of Physics,*** Dover Publications, NewYork NY.

Morse, A.S., (1990), "Towards a Unified Theory of Parameter Adaptive Control: Tunability", *IEEE Trans. on Automatic Control*, Vol. 35, No. 9, pp. 1002-1012, September.

Morse, A.S., (1992), "Towards a Unified Theory of Parameter Adaptive Control-Part II:Certainty Equivalence and Implicit Tuning", *IEEE Trans. on Automatic Control*, Vol. 37, No. 1, pp. 15-29, January.

Prigogine, Ilya, (1980), ***From Being to Becoming***, Freeman and Company, San Francisco.

Prigogine, Ilya, (1996), ***La Fin des Certitudes*** Editions Odile Jacob, Paris France.

Rifkin, Jeremy, (1989) ***Entropy into the Greenhouse World*** Bantam Books New York.

Saridis, G.N. (1985), "An Integrated Theory of Intelligent Machines by Expressing the Control Performance as an Entropy", *Control Theory and Advanced Technology*, **Vol. 1**, No. 2, pp. 125-138, Aug.

Saridis, G.N. (1988), "Entropy Formulation for Optimal and Adaptive Control", *IEEE Transactions on Automatic Control*, **Vol. 33**, No. 8, pp. 713-721, Aug.

Saridis, G.N. (1989), "Analytic Formulation of the IPDI for Intelligent Machines", *AUTOMATICA the IFAC Journal,* ***25***, No. 3, pp. 461-467.

Saridis, G. N. (1995) ***Stochastic Processes, Estimation, and Control: The Entropy Approach*****,** John Wiley and Sons, New York.

Saridis, G.N. and Lee, C.S.G. (1979), "Approximation of Optimal Control for Trainable Manipulators", *IEEE Trans. on Systems Man and Cybernetics*, **Vol.8**, No. 3, pp. 152-159, March.

Shannon, C., Weaver W. (1963) ***The Mathematical Theory of Communications***, Aeolian Books, Urbana Ill.

Singh, Madan G., (editor), (1987) ***Systems and Control Encyclopedia; Theory, Technology Applications***, Vol. 1-8, Pergamon Press, Oxford UK.

Zames G. (1979), "On the metric complexity of causal linear systems, ε- entropy and ε-dimension for continuous time" *IEEE Transactions on Automatic Control*, **Vol. AC-124**, pp. 222-230.

CHAPTER 2

STOCHASTIC OPTIMAL ESTIMATION AND CONTROL

2.1 INTRODUCTION

In Chapter 1, entropy was established as the cost of producing a control action on a system. The equivalence between the two of them was even established analytically in section 1.6. In this Chapter, the derivation of the optimal deterministic, stochastic, both continuous and discrete-time, suboptimal, adaptive, and dual control will be obtained through entropy generated analytic expressions. The same approach will be followed for continuous and discrete-time state estimation.

Entropy will generate the formulation of the control problem for a given but arbitrary controller lying away from the optimal/equilibrium value, in the sense of chaos theory, as in section 1.6. With this formulation, suboptimal iterative solutions will also be generated which converge to the optimal as in Saridis and Lee (1979).

The claim of the Dual control problem, introduced by Feld'baum (1965), will be verified analytically, using entropy, deriving the truly optimal control for nonlinear systems. This will revoke the claim that the "Certainty Equivalence Principle" is not optimal but only adaptive. It must be made clear that adaptive control of a system completely described analytically, makes sense only in stochastic environments in spite of other opinions (Morse 1990,1992). Typical examples will demonstrate these statements.

2.2 THE DETERMINISTIC OPTIMAL CONTROL

The entropy formulation of the control problem, as shown in section 1.6, yields the same results as the classical optimal control theory by minimizing the Generalized Hamilton-Jacobi-Bellman equation. However, it has the advantage of connecting this theory to other systems using Entropy as a measure of performance.

Furthermore, the deterministic version of the Theory of Nonlinear Optimal Control may be obtained from the properties of the density function p(u) by eliminating the stochastic variables w(t), v(t), and x(0);

$$dx/dt = f[x,u,t]\ ; \qquad x(t_0) = x_0\ ;\ p(x_0)$$
$$z(t) = h(x,t) + v(t); \tag{2.1}$$

then

$$p(u) = e^{-\lambda-\mu V(u(x,t),x0,t0)} \tag{2.2}$$

$$e^{\lambda} = \int_{\Omega x} e^{-\mu V(u(x,t),x0,t0)}\, dx \tag{2.3}$$

This density must satisfy "the time incompressibility condition", of probability functions, equivalent to the Liouville equation (Prigogine 1980).

$$dp/dt = 0 \Leftrightarrow [\partial p/\partial t + \partial p^T/\partial x \cdot dx/dt + L(x,u,t)]p(u) = 0 \tag{2.4}$$

where $\partial p/\partial t = -\partial V/\partial t \cdot p(u)$, $\partial p(u)/\partial x = -\partial V/\partial x \cdot p(u)$, and $dx/dt = f(x,u,t)$ for some given u{x,t). Then,

$$[\partial V/\partial t + \partial V^T/\partial x \cdot f(x,u,t) + L(x,u,t)]p(u) = 0$$

Since, eq. (2.2) implies that $p(u) \neq 0$, for all $x \varepsilon \Omega_x$;

$$[\partial V/\partial t + \partial V^T/\partial x \cdot f(x,u,t) + L(x,u,t)]p(u) = 0;$$

$$V(X(T),T) = 0, \tag{2.5}$$

which is the **Generalized Hamilton-Jacobi** equation for an arbitrary admissible control, derived by Saridis and Lee in "Approximation Theory of Optimal Control" (1979).

Minimization of eq. (2.5) yields the optimal control that minimizes the differential entropy H(u), e.g.;

$$\partial V/\partial t + \min_{u} [\partial V^T/\partial x \cdot f(x,u,t) + L(x,u,t)]p(u) = 0;$$

$$V(x(T),T) = 0, \tag{2.6}$$

The following theorem summarizes the above results.

THEOREM 2.1: A necessary and sufficient condition for $u^*(x,t)$ to minimize $V(u(x,t),x_0,t_0)$, subject to the system equations,

$$dx/dt = f(x,u,t); \qquad x(t_0) = x_0,$$

is the $u^*(x,t)$ that minimizes the differential entropy H(u,p(u)), where p(u) is the worst entropy density function according to Jaynes' Maximum Entropy Principle.

2.3 THE STOCHASTIC OPTIMAL CONTROL PROBLEM

The stochastic version of the Theorem, through the entropy reformulation, is obtained the same way as the deterministic case.

Assume the following stochastic system, and performance cost are given.
The System Equations are:

$$dx/dt = f[x,u,t] + g(x,t)w(t); \qquad x(t_0) = x_0 \; ; \; p(x_0) \tag{2.7}$$
$$z(t) = h(x,t) + v(t);$$

where the random processes, and their associated probability densities, w(t) ; p(w) : v(t) ; p(v), and their initial conditions (eq. 1.12) are given.

The Performance Cost is:

$$V[u(t),x_0,t_0] = E\{J[x_0,t_0,u(x,t),w(t),v(t)]\} = E\{\int_{t0}^{T} L(x,u,t)\, dt/x_0\} \tag{2.8}$$

The p(u(x,t)) is derived as Jaynes' Worst Entropy Probability Density Function:

$$p(u(x,t)) = \exp[-\lambda - \mu E\{V[u(x,t),x_0,t_0]\}] \tag{2.9}$$

Then the Associated Entropy is:

$$H[u(x,t)] = \lambda + \mu E\{V[u(x),x_0,t_0]\}. \tag{2.10}$$

which may replace the Performance Criterion for Optimization.

Jaynes' Principle of Maximum Entropy implies:

$$dH/dp = 0.$$

The Incompressibility in time of Probability Density Condition implies:

$$dp/dt = 0$$

which for the selected p(u(x,t)) and a fixed but arbitrary control u(x,t), yields the **Generalized Hamilton-Jacobi-Bellman** equation, derived in section 1.6:

$$\partial V/\partial t + \partial V^T/\partial x \cdot f(x,u,t) + L(x,u,t) + \tfrac{1}{2}\mathcal{L}\{Tr(\partial V^2/\partial x^2 Qgg^T)\} = 0$$

$$V(T) = 0. \tag{2.11}$$

Minimization of eq.(6.25) yields again the stochastic optimal control

$$\partial V/\partial t + \mathrm{Min}_u\, [\partial V^T/\partial x \cdot f(x,u,t) + L(x,u,t) + \tfrac{1}{2}\mathcal{L}\{Tr(\partial V^2/\partial x^2 Qgg^T)\}] = 0;$$

$$V(x(T),T) = 0. \tag{2.12}$$

The following theorem establishes the equivalence of the stochastic optimal feedback control problem with the minimization of the worst case differential entropy.

.**THEOREM 2.2:** A necessary and sufficient condition for u(x,t) to minimize

$$V[u(x,t),x_0,t_0] = E\{J(u(x,t),x_0,t_0)\}$$

subject to the dynamic constraints:

$$\begin{aligned} dx/dt &= f(x,u,t) + g(x,t)w(t); \quad x(t_0) = x_0: \\ z(t) &= h(x,t) + v(t); \end{aligned} \tag{2.13}$$

where w(t), v(t), are i.i.d random processes; is that u(x,t) minimizes the **differential entropy H[u(x,t)]**, where the associated p(u(x,t)) is the **maximum entropy density function**, satisfying **Jaynes' Maximum Entropy Principle**

The proof of the theorem is given in Saridis (1988).

The above theorem may be used to obtain the solution of the stochastic optimal control problem. It may also be extended to define other design problems as Minimum Entropy problems, e.g., the parameter identification, optimal state estimation and dual control problems. It also serves in establishing analytically, Fel'dbaum's claim that the **feedback active optimal control** is really the globally Minimal solution of the stochastic optimal control problem.

2.4 THE STOCHASTIC SUBOPTIMAL CONTROL PROBLEM

The Suboptimal Control problem is a result of the Approximation Theory developed by Saridis and Lee (1979),with a new fancy name. It deals with the general control problem of the behavior of a system under the control of an arbitrary but fixed controller, which may lie away of the optimal value considered to be the equilibrium. It may be reformulated using Jaynes's Principle of Maximum Entropy.

The Approximation Theory for stochastic systems claims that the performance cost function V(u(x);x,t) satisfies the Generalized Hamilton-Jacobi-Bellman equation for some admissible control $u(x) = u_i(x)$, I=1,2,...

Such a Generalized Hamilton-Jacobi-Bellman equation was derived in section 1.6, for some control u(x,t), using Jaynes's Principle of Maximum Entropy. To complete the claim one may assume that $u(x,t) = u_i(x)$ in eq.(2.12). Then the following theorem establishes the claim that for a certain algorithm a sequence of generated control converges to the optimal.

THEOREM 2.3: The performance cost $V(u_i(x);x,t)$, of system (2.13) for an admissible control $u_i(x)$, may be expressed by its associated Entropy $H(u_i(x,t), I=1,..,$ derived by Jaynes's Principle of Maximum Entropy.

$$u_i(x) = -\tfrac{1}{2} b^T \partial V_{i-1}/\partial x \quad \Rightarrow \quad V_{i-1} \geq V_i \tag{2.13}$$

Lemma 2.1 For $I \to \infty$ the control $u_i(x)$ converges to the optimal control $u^*(x)$.

The derivation of the rest of the theorems of the Approximation Theory of stochastic systems follow the same procedure as in (Saridis 1995). Their proofs are also presented in Saridis (1988). This includes the Upper and Lower Bounds of the Approximation, the Infinite-time optimal control problem, and the design procedures.

2.5 DISCRETE-TIME FORMULATION OF THE STOCHASTIC OPTIMAL CONTROL PROBLEM

The discrete-time stochastic optimal control problem may be reformulated as an Entropy minimization problem, associated with the worst case probability, in a way similar to the continuous-time case presented in section 2.3. The Entropy approach to include both deterministic and stochastic discrete-time optimal control problems was presented in (Tsai, Casiello, Loparo, 1992). The stochastic version of the Entropy formulation of the discrete-time stochastic optimal control problem is presented here.

Assume that the following discrete-time stochastic system, performance cost, the random processes $w(k)$, $v(k)$ and their respective associated probability densities $p(w)$, $p(v)$, are given.

The System Equations are:

$$\begin{aligned} x(k+1) &= f(x(k),u(k),k) + g(x(k),k))w(k); \qquad x(0) = x_0\ ;\ p(x_0) \\ z(k) &= h(x(k),k) + v(k); \end{aligned} \tag{2.14}$$

where $u(k) = u(x,k) \in \Omega_u$, the set of admissible controls.

The Performance Cost is:

$$V_N[u(x,k),x_0] = E\{J[x_0,u(x,k),w(k),v(k)/x_0\} = E\{\sum_{k=0}^{N-1} L(x(k+1),u(k),k)/x_0\} \tag{2.15}$$

Derive $p(u(x,k))$, as Jaynes' Worst Entropy Probability Density Function:

$$p(u(x,k)) = \exp[-\lambda - \mu E\{V[u(x,k),x_0]\}] \qquad (2.16)$$

Then the associated Entropy is:

$$H[u(x,k)] = -\int_{\Omega x} p(u(x))\ln p(u(x))\,dx = \lambda + \mu E\{V[u(x,k),x_0]\}. \qquad (2.17)$$

which may replace the Performance Criterion for Optimization.

The worst case density function according to Jaynes' Principle of Maximum Entropy was obtained from:

$$dH/dp = 0. \qquad (2.18)$$

The Bellman equation that governs the dynamics of the discrete-time system is given, for an arbitrary but fixed admissible controller $\{u_i(k)\}$, used in the approximation theory, by,

$$V_{N-k}(x(k)/Z^k) = E\{L(x(k+1),u_i(k)) + V_{N-k-1}(x(k+1)/Z^{k+1})/Z^k\} \qquad (2.19)$$

and for the optimal controller $u^*(k)$;

$$V_{N-k}(x(k)/Z^k) = \operatorname*{Min}_{u} E\{L(x(k+1),u_i(k)) + V_{N-k-1}(x(k+1)/Z^{k+1})/Z^k\} \qquad (2.20)$$

Minimization of eq.(2.17) yields again the stochastic optimal control for the discrete-time problem.

THEOREM 2.4: A necessary and sufficient condition for u(x,t) to minimize

$$V[u(x,k),x_0] = E\{J(u(x,k),x_0)\}$$

subject to the dynamic constraints:

$$x(k+1) = f(x(k),u(k),k) + g(x(k),k)w(k); \qquad x(0) = x_0\ ;\ p(x_0)$$
$$z(k) = h(x(k),k) + v(k);$$

where w(k), v(k), are i.i.d random processes; is that u(x,k) minimizes the **differential entropy H[u(x,k)]**, where the associated p(u(x,k)) is the **maximum entropy density function**, satisfying **Jaynes' Maximum Entropy Principle**.

The above theorem may be used to obtain the solution of the discrete-time stochastic optimal control problem. It may also be extended to define other design problems as Minimum Entropy problems, e.g., the parameter identification, optimal state estimation and dual control problems. It also serves in establishing analytically, Fel'dbaum's claim that the **feedback active optimal control** is really the globally Minimal solution of the stochastic optimal control problem.

2.6 MAXIMUM ENTROPY FORMULATION OF STATE ESTIMATION: CONTINUOUS-TIME

State Estimation for continuous-time and discrete-time systems can be easily formulated, using Entropy measures, and Jaynes' Principle of Maximum Entropy.

The continuous-time state estimation problem both in the linear case, e.g. the Kalman-Bucy filter (Fig.2.1), as well as the nonlinear case, e.g. the MMSE estimator, may be reformulated as a Maximum Entropy problem, by using Jaynes' principle of **Maximum Entropy**. Following a procedure similar to the optimal control reformulation for the continuous-time system, eq.(2.8);

$$\begin{aligned} dx/dt &= f(x,u,t) + g(x,t)w(t);\; x(t_0) = x_0\,;\; p(x_0) \\ z(t) &= h(x,t) + v(t); \qquad w(t);p(w):v(t);p(v) \end{aligned} \tag{2.21}$$

and the Risk function, for $t \in \mathfrak{J} = [t_0,T]$;

$$R'(\hat{x}) = E\{\int_{\mathfrak{J}} (\hat{x}-x)^T A(\hat{x}-x)\, dt\} \tag{2.22}$$

where $A = \Lambda^{-1}$ the inverse of the measurement noise covariance matrix. It is easy to find the worst case entropy according to Jaynes' principle.

$$\begin{aligned} I' &= \beta H(\hat{x}) - \gamma[E\{\int_{\mathfrak{J}} \|\hat{x}-x\|_A^2\, dt\} - K] - \alpha\left[\int_{\Omega x} p(\hat{x})d\hat{x} - 1\right] \\ &= -\int_{\Omega\hat{x}} [\beta p(\hat{x})\ln p(\hat{x}) + \gamma p(\hat{x})\int_{\mathfrak{J}}\|\hat{x}-x\|_A^2 dt]d\hat{x} - \alpha\left[\int_{\Omega x} p(\hat{x})d\hat{x} - 1\right] \end{aligned} \tag{2.23}$$

Using the Lemma of Calculus of Variations, maximization of I with respect to p(u) yields,

$$\partial[-\beta p(\hat{x})\ln p(\hat{x}) - \gamma p(\hat{x})\int_{\mathfrak{J}}\|\hat{x}-x\|_A^2 dt - \alpha p(\hat{x})]/\partial p = 0; \qquad \partial^2 I/\partial p^2 < 0$$

or

$$-\beta\ln p(\hat{x}) - \beta - \gamma\int_{\mathfrak{J}}\|\hat{x}-x\|_A^2 dt - \alpha = 0; \qquad -\beta/p(\hat{x}) < 0$$

and the worst case density is,

$$\begin{aligned} p(\hat{x}) &= \exp\{-\nu-\rho\int_{\mathfrak{J}}\|\hat{x}-x\|_A^2 dt\} \\ e^{\lambda} &= \int_{\Omega\hat{x}} \exp\{-\rho\int_{\mathfrak{J}}\|\hat{x}-x\|_A^2 dt\}\, dx \end{aligned} \tag{2.24}$$

with $\nu = \gamma/\beta$, $\rho = (\alpha+1)/\beta$, and

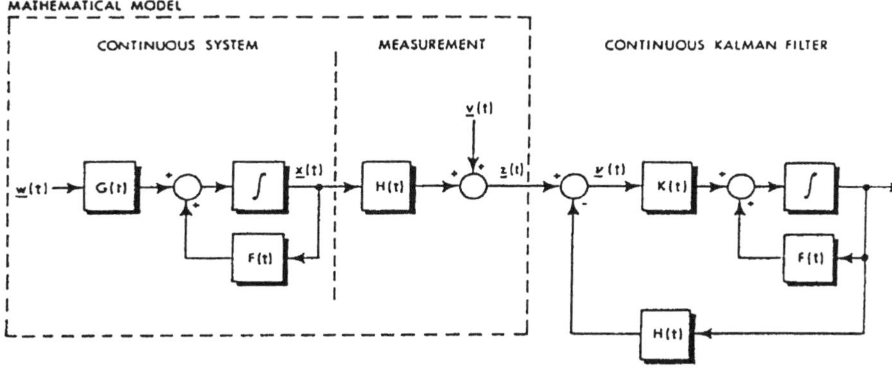

Fig. 2.1 System Model and Continuous-time Kalman-Bucy Filter

$$R'(\hat{x}) = K = \partial/\partial v \ln \int_{\Omega\hat{x}} \exp\{-v\int_{\mathfrak{T}} \|\hat{x}-x\|_{A}^{2} dt\}\, dx \tag{2.25}$$

The associated Worst Case Entropy then is given by;

$$H(\hat{x}) = \rho + vE\{\int_{\mathfrak{T}} [\hat{x}-x]^{T} \Lambda^{-1} [\hat{x}-x]\, dt\} \tag{2.26}$$

and minimization of $H(\hat{x})$ yields the optimal MMSE. (minimum mean square error) estimator. The following theorem is a result of the previous formulation:

THEOREM 2.5: The optimal MMSE. state estimator for the system of eq. (2.22) is obtained by minimizing the differential entropy $H(\hat{x})$, where the associated $p(\hat{x})$ is **the maximum entropy density function** satisfying **Jaynes' Maximum Entropy Principle.**

In the case of a linear gaussian system, the MDSE. estimator gives the most probable estimator as well, according to Bayes' theorem. This problem will be considered in the sequel as part of the entropy formulation of the Linear Quadratic Gaussian (LQG) optimal control problem.

Alternate formulations of the optimal state estimation problem, using entropy criteria are also available in the literature (Kalata, Premier 1974).

2.7 MAXIMUM ENTROPY FORMULATION OF STATE ESTIMATION: DISCRETE-TIME.

The problem of discrete-time State Estimation is also a direct extension of the continuous-time estimation problem discussed in the previous section. This problem was also treated by Loparo and his colleagues in (Tsai, Casiello, Loparo, 1992), and is a generalization of the Discrete Kalman Filter (Fig. 2.2). The results of the Entropy formulation are presented hereafter.

Assume that the following discrete-time stochastic system, performance cost, the random processes w(k), v(k) and their associated probability densities p(w), p(v), are given.

Given the system equations,

$$\begin{aligned} x(k+1) &= f[x(k),u(k),k] + g[x(k),k)w(k); \qquad x(0) = x_0\ ;\ p(x_0) \\ z(k) &= h[x(k),k) + v(k); \end{aligned} \tag{2.27}$$

where $u(k) = u(x,k) \in \Omega_u$, the set of admissible controls, and the Risk function, for $k \in \mathcal{H} = [0,N]$;

$$R'(\hat{x}) = E\{\sum_{k=0}^{N-1} [\hat{x}(k)-x(k)]^{T} \Lambda^{-1} [\hat{x}(k)-x(k)]/Z^{N-1}\} \tag{2.28}$$

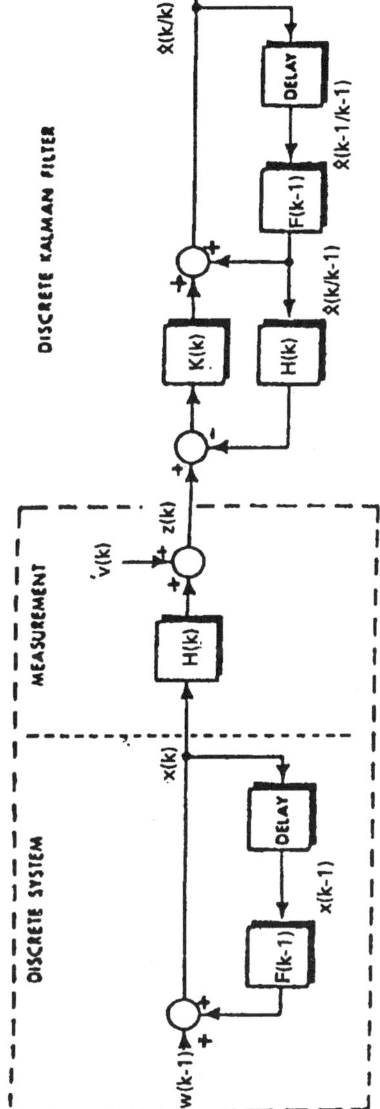

Fig. 2.2 System Model and Discrete-time Kalman Filter

The worst case Entropy according to Jaynes' principle is found in a similar way as previously, and the worst case density function is,

$$p(\hat{x}) = \exp[-\lambda-\mu E\{\sum_{k=0}^{N-1} [\hat{x}(k)-x(k)]^T\Lambda^{-1}[\hat{x}(k)-x(k)]/Z^{N-1}\}] \tag{2.29}$$

$$e^{\lambda} = \int_{\Omega\hat{x}} \exp\{-\rho\sum_{k=0}^{N-1}\|\hat{x}-x\|_A^2\}\, dx$$

and

$$R'(\hat{x}) = \partial/\partial v \ln\int_{\Omega\hat{x}} \exp\{-v\sum_{k=0}^{N-1}\|\hat{x}-x\|_A^2\}\, dx \tag{2.30}$$

The associated Worst Case Entropy then is given by;

$$H(\hat{x}) = \rho + vE\{\sum_{k=0}^{N-1} [\hat{x}(k)-x(k)]^T\Lambda^{-1}[\hat{x}(k)-x(k)]/Z^{N-1}\} \tag{2.31}$$

and minimization of $H(\hat{x})$ yields the optimal MDSE. (minimum risk) estimator. The following theorem is a result of the above formulation:

THEOREM 2.6: The optimal MDSE. state estimator for the system of eq.(2.20) is obtained by minimizing the differential entropy $H(\hat{x})$, where the associated $p(\hat{x})$ is **the maximum entropy density function** satisfying **Jaynes' Maximum Entropy Principle.**

In the case of a linear gaussian system, the MDSE. estimator gives the most probable estimator as well, according to Bayes' theorem. This problem will be considered in the sequel as part of the entropy formulation of the Linear Quadratic Gaussian (LQG) optimal control problem (Fig. 2.3).

2.8 THE COST OF ACTIVE FEEDBACK (DUAL) CONTROL PROBLEM

As it is well known "active feedback (dual) control", developed through dynamic programming represents the optimal solution for stochastic nonlinear systems. Other solutions that use feedback control until the present time and predictive open loop control from there on, based on the **Certainty Equivalence Principle**, and known as **open loop feedback optimal** are claimed not to be optimal.

> **The Stochastic Optimal or Adaptive Control have the same form as if the Estimate of the state and/or the unknown parameter were certain, without error.**

Most of the existing stochastic Open-Loop-Feedback-Optimal and Adaptive Control

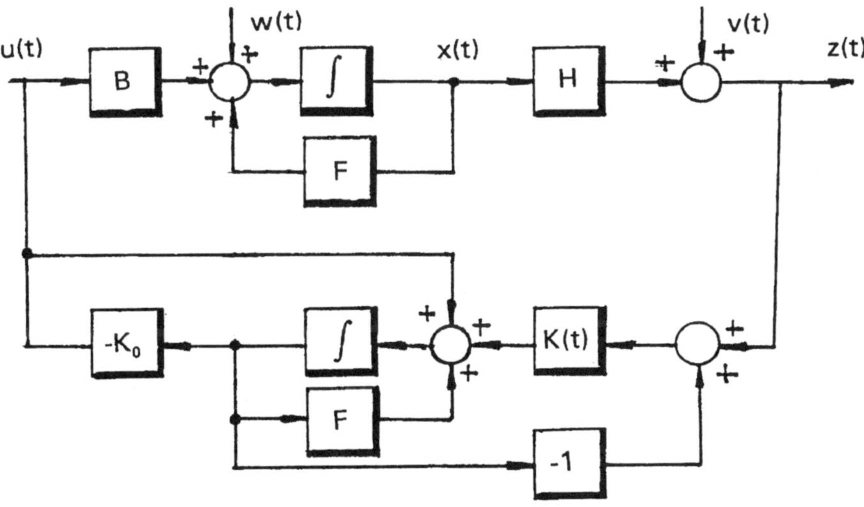

Fig. 2.3 Continuous-time Stochastic Optimal LQG Control

algorithms have been derived based on this Principle [Kumar and Varaiya 1986]:

The case of active feedback (dual) control, presented by Feld'baum (1965) as the only optimal control for nonlinear stochastic systems, was with no mathematical verification, due to the lack of analytic tools at that time. The entropy reformulation of the stochastic optimal control problem provides those tools through the partition properties of the differential entropy of the control and state estimation activities. The cases of stochastic optimal control and stochastic optimal adaptive control will be considered here, and thus analytically verify Fel'dbaum's claim (1965), that only **the active feedback optimal control** is the globally optimal solution.

This principle is correct only for neutral systems like in the Linear Quadratic Gaussian (LQG} case. This is analytically shown in the sequel.

However, the Certainty Equivalence Principle is useful in designing adaptive controls when the estimation and/or parameter identification process converges fast enough. In this section, it will be shown through the entropy partition theorem, that such solutions are not optimal but one may establish measures of deviation from optimality.

2.9 STOCHASTIC OPTIMAL (DUAL) ESTIMATION AND CONTROL

The case of simultaneous optimal estimation and control for linear systems with gaussian noise and quadratic cost functions, which belong to the category of neutral systems, and therefore yield globally optimal solutions by cascading a Kalman-Bucy filter with a deterministic optimal control. This result misled people to generalize the concept for non-neutral cases and claim the Certainty Equivalence Principle as optimal. Optimal estimation is necessary since the states are not readily available for feedback.

The problem of non-neutral stochastic nonlinear control, which includes the case of Adaptive control, named by Feld'baum as **Dual,** and which is the truly optimal solution, is now recast using entropy as the cost of the operation.

Consider the worst case differential entropy of the optimal control $H(u^*)$. It may be partitioned as follows, using the definition of marginal probabilities, to take into account the underlying state estimation problem:

$$H(u^*) = -\int_{\Omega x} p(u^*)\ln p(u^*)\,dx = -\int_{\Omega x}\int_{\Omega\hat{x}} p(u^*,x)\ln p(u^*)\,d\hat{x}dx =$$

$$= -\int_{\Omega x}\int_{\Omega\hat{x}} p(u^*/\hat{x})p(\hat{x})\ln[p(u^*/\hat{x})p(\hat{x})/p(\hat{x}/u^*)]\,d\hat{x}dx =$$

$$= -\int_{\Omega x}\left[\int_{\Omega\hat{x}} p(u^*/\hat{x})p(\hat{x})\ln[p(u^*/\hat{x})]dx\right]p(x)d\hat{x} -$$

$$-\int_{\Omega x}\left[\int_{\Omega\hat{x}} p(u^*/\hat{x})dx\right]\ln p(\hat{x})d\hat{x} - \int_{\Omega x}\left[\int_{\Omega\hat{x}} p(\hat{x}/u^*)\ln p(\hat{x}/u^*)d\hat{x}\right]p(u^*)dx$$

The final expression may be rewritten in terms of differential entropies as;

$$\mathbf{H[u^*] = H[u^*/\hat{x}] + H[\hat{x}] - H[\hat{x}/u^*]} \tag{2.32}$$

The decomposition (2.32), explained by the following theorem, justifies Fel'dbaum's claim, that the active feedback control accounting for the control function when estimating future states, is **globally minimum**. It also disclaims that the partition theorem of stochastic control is optimal in all cases outside the LQG.

THEOREM 2.7: The cost of optimal control with inaccessible states is the cost of control with estimates of the states x, plus the cost of $\hat{x}$, plus the cost of state estimation, minus the cost of active transmission of information.

2.10 STOCHASTIC SUBOPTIMAL CONTROL REVISITED

The generalized theory of suboptimal control, providing solutions even away from the optimal solution, discussed in section 2.4, accepts the same treatment as the Dual control.

A similar decomposition and a suitable theorem, applies to the suboptimal control.

$$H[u_i] = H[u_i/\hat{x}] + H[\hat{x}] - H[\hat{x}/u_i] \tag{2.33}$$

THEOREM 2.8: The cost of suboptimal control u_i with inaccessible states is the cost of control with estimates of the states x, plus the cost of $\hat{x}$, plus the cost of state estimation, minus the cost of active transmission of information for control u_i.

2.11 STOCHASTIC OPTIMAL ADAPTIVE CONTROL

The **Optimal Adaptive Control** problem evolves from the Stochastic Optimal Control problem when the system parameters $\varphi\varepsilon\Omega_{\varphi}$, are unknown or unaccessible for measurement. The system may be described by;

$$dx/dt = f(x,u,\varphi,t) + g(x,\varphi,t)w(t); \quad x(t_0) = x_0\ ;\ p(x_0) \tag{2.34}$$
$$z(t) = h(x,t) + v(t); \qquad w(t);p(w):v(t);p(v)$$

and it is required that;

$$H(u^*) = \mathrm{Min}_u H(u) \Leftrightarrow \mathrm{Min}_u(x_0,u,t) \tag{2.35}$$

subject to optimal state estimation, and parameter identification. Assuming a Least Squares algorithm for parameter identification φ, a partition of the differential entropy of the stochastic system, $H(u^*)$, similar to the **Stochastic Optimal Control** problem, yields:

$$H[u^*] = H[u^*/\hat{x},\hat{\phi}] + H[\hat{x}/\hat{\phi}] + H[\hat{\phi}] - H[\hat{x},\hat{\phi}/u^*] \tag{2.36}$$

where $H[\hat{x},\hat{\phi}/u^*]$ is the **Active Transmission of Information** both in State Estimation $\hat{x}$, and Parameter Identification $\hat{\phi}$.

A theorem similar to 2.7, may be stated for the case of the optimal Adaptive Control:

THEOREM 2.9: The cost of control with inaccessible states and parameters is the cost of control with estimates of the states $\hat{x}$ and parameters $\hat{\phi}$, plus the cost of state estimation with estimates of parameters $\hat{\phi}$, minus the cost of active transmission of information of the states and parameters.

Typical Adaptive Control algorithms (Saridis 1977), (Kumar and Varaiya 1986), (Morse 1992) are based on the **Certainty Equivalence Principle**, which take into consideration only the cost control using the state and parameter estimates and that of the state estimation and parameter identification. It is clear from the above formulation that these algorithms are **definitely not optimal**.

This is demonstrated by the following inequality:

$$H[u]_{ad} = H[u^*/\hat{x},\hat{\phi}] + H[\hat{x}/\hat{\phi}] + H[\hat{\phi}] = H(u^*) + H[\hat{x},\hat{\phi}/u^*] > H(u^*) \tag{2.37}$$

The following example demonstrates this statement, which agrees with Fel'dbaum's claim of Dual Control.

2.11.1 EXAMPLE: THE DUAL-OPTIMAL AND ADAPTIVE CONTROL

The difference between the Dual-optimal and Adaptive Control is demonstrated with the comparison of an Adaptive to an LQG Optimal Control problem. Given the system:

$$dx_1/dt = x_2 \qquad x(0) \sim N(0,10I)$$
$$dx_2/dt = \varphi x_2 + u + w \qquad \varphi = -1$$
$$z = x_1 + v \qquad w(t) \sim (0,1)\ ;\ v(t) \sim (0,1)$$

and the performance cost;

$$J(u) = E\{\int_0^T (2x_1^2 + 2x_2^2 + \tfrac{1}{2}u^2)\, dt\}$$

The problem is to calculate the cost function J for the case of φ been unknown

a. the adaptive solution, obtained from the **Principle Certainty Equivalence**

and

b. the case that φ is known e.g. the optimal solution.

a. The adaptive case

$$H(u^*,\hat{x},\phi) = J(x(0),u^*(\hat{x},\phi)) = E\{[9+\phi+\phi^2+(\phi+1)\sqrt{\phi^2+8}]x(0)_1{}^2/$$
$$/[(\phi+1)+\sqrt{\phi^2+8}]+2x(0)_1x(0)_2+[4+\tfrac{1}{2}\phi^2+\tfrac{1}{2}\phi\sqrt{\phi^2+8}]x(0)_2{}^2$$
$$/[(\phi+1)+\sqrt{\phi^2+8}]\}$$

with

$$u^*(x) = -2\hat{x}_1 - [\phi + \sqrt{\phi^2+8}]\, x(0)_2$$

$$H(\hat{x}/\phi) = \int_0^T TrP(t)\, dt = \int_0^T (p(t)_{11} + p(t)_{22})\, dt$$

where $P(t) = \begin{bmatrix} p_{11} & p_{12} \\ p_{12} & p_{22} \end{bmatrix}$ $P(0) = 10I$, the error covariance matrix and

$$dp_{11}/dt = 2p_{12} - p_{11}{}^2 \qquad p(0)_{11} = 10$$
$$dp_{12}/dt = p_{22} + \varphi p_{12} - p_{11}p_{12} \qquad p(0)_{22} = 10$$
$$dp_{22}/dt = 2\varphi p_{22} - p_{12}{}^2 \qquad p(0)_{12} = 0$$

and x is estimated by a Kalman filter;

$$d\hat{x}_1/dt = \hat{x}_2 + p_{11}(z - \hat{x}_1)$$
$$d\hat{x}_2/dt = \phi\hat{x}_1 + p_{12}(z - \hat{x}_1)$$

The parameter estimation is performed with a Least Squares Identification algorithm (Saridis 1977), after time discretization;

$$\psi(k+1) = \psi(k) + S(k+1)\omega(k)[\omega(k+1) - T^2\omega(k)]$$
$$S(k+1) = S(k) - S(k)^2\omega(k)^2[\omega(k)S(k)\omega(k) + 1]^{-1}$$
$$\omega(k) = z(k) - z(k-1)$$
$$\psi(k) = 1 + \phi T; \qquad T = 0.1 \text{ sec.}$$

The results of the state estimation are given in Fig. 2.4, and the results of the parameter identification in Fig. 2.5. The latter have been used in the state estimation algorithm. The controller $u^*(\hat{x},\phi)$, uses the results of both estimation and identification algorithms and are depicted in Fig. 2.6. Fig. 2.7 depicts the Cost functions.

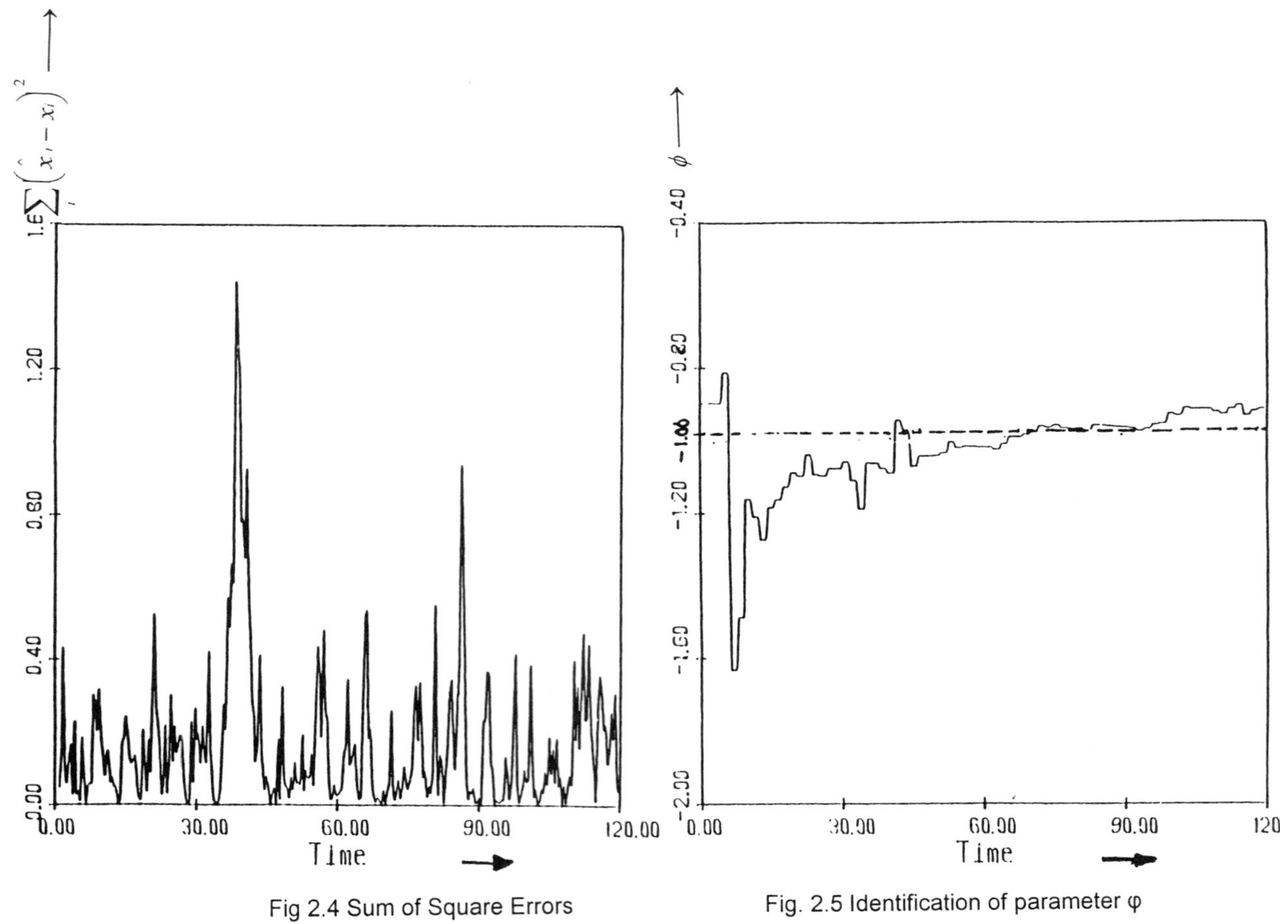

Fig 2.4 Sum of Square Errors

Fig. 2.5 Identification of parameter φ

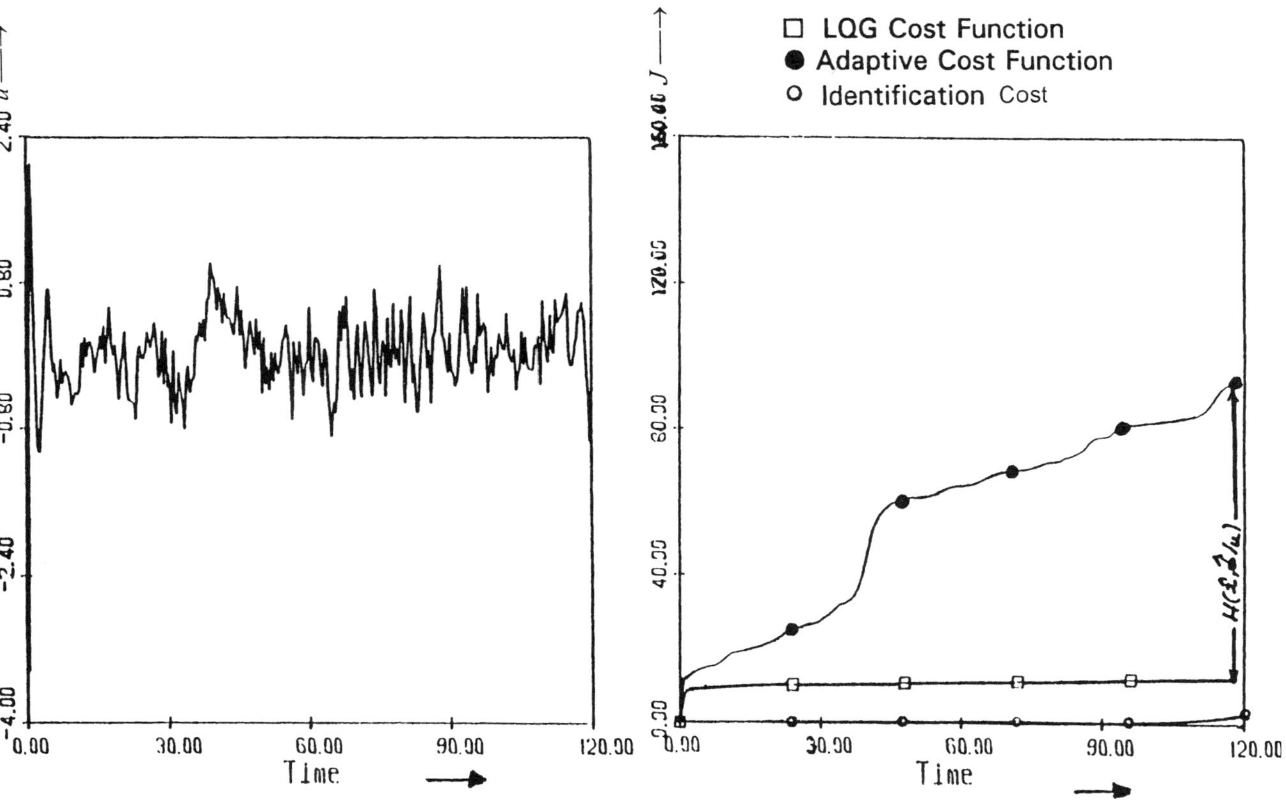

Fig. 2.6 Control Input

Fig. 2.7 Cost Functions

b. The Optimal Linear Quadratic Gaussian case

The optimal Linear Quadratic Gaussian case (LQG) solution is obtained by the following steps;

$$H(u/\hat{x}) = J(x_0,u(\hat{x})) = E\{\int_0^T (2\hat{x}_1^2 + 2\hat{x}_2^2 + \tfrac{1}{2}u^2)\,dt\}$$

and the optimal controller is;

$$u^*(\hat{x}) = -2\hat{x}_1 - 2\hat{x}_2$$

The differential entropy of estimation $H(\hat{x})$, is obtained from the state estimates of the Kalman filter equations;

$$d\hat{x}_1/dt = \hat{x}_2 + p_{11}(z - \hat{x}_1)$$
$$d\hat{x}_2/dt = \phi\hat{x}_1 + p_{12}(z - \hat{x}_1)$$
$$dp_{11}/dt = 2p_{12} - p_{11}^2 \qquad p(0)_{11} = 10$$
$$dp_{12}/dt = p_{22} + \varphi p_{12} - p_{11}p_{12} \qquad p(0)_{22} = 10$$
$$dp_{22}/dt = 2\varphi p_{22} - p_{12}^2 \qquad p(0)_{12} = 0$$

$$H(\hat{x}) = \int_0^T \mathrm{Tr}P(t)\,dt = \int_0^T (p(t)_{11} + p(t)_{22})\,dt$$

Then the Dual control cost is given by;

$$H(u^*) = H(u^*/\hat{x}) + H(\hat{x})$$

Now, for comparison purposes, the cost of Identification $H(\phi)$ is removed from the cost of Adaptive control. This cost is calculated from the results of Fig. 2.7:

$$H(\phi) = \sum_{l=0}^{k} (\varphi_i - \phi_l)^2$$

The resulting Adaptive control cost, is compared to the Dual control cost in Fig.2.7;

$$H(u/\phi) = H(u/\hat{x},\phi) + H(\hat{x}/\phi) > H(u^*)$$

The difference between the two is the **equivocation of active transmission of information**, and as expected is positive which proves that the Adaptive control is not optimal.

This difference could serve as a measure of closeness of the Adaptive to the Dual control solution.

2.12 THE LQG OPTIMAL CONTROL AND THE KALMAN-BUCY FILTER

The difficulties arising from nonlinear optimal estimation and control are not present in the case of Neutral Systems, like Linear Quadratic Gaussian (LQG) systems given by;

$$dx/dt = F(t)x(t) + B(t)u(t) + G(t)w(t); \qquad x(0) \sim N(x_0,P_0) \tag{2.38}$$
$$z(t) = H(t)x(t) + v(t) \qquad w(t)\sim N(0,Q);v(t)\sim N(0,R)$$
$$J(u) = E\left\{\int_0^T [x^TMx + u^TNu]\,dt\right\} = H(u) = E\{I(u)\}.$$

Selecting p(x) to correspond to the Minimum Mean Square Error estimator eq. (2.24), the Kalman-Bucy filter is derived from the associated worst case differential entropy $H(\hat{x})$;

$$H(\hat{x}) = -\int_{\Omega x} p(\hat{x})\ln p(\hat{x})\,dx = \rho + v\int_{\Omega x}\int_0^T \|\hat{x}-x\|^2\,dt\, p(\hat{x})dx =$$

$$= \rho + vE\left\{\int_0^T \|\hat{x}-x\|^2\,dt\right\} \tag{2.39}$$

Minimization of the differential entropy implies;

$$\underset{x}{\text{Min}}\; H(\hat{x}) \Leftrightarrow \underset{x}{\text{Min}}\; E\left\{\int_0^T [\|\hat{x}-x\|^2 + \lambda^{TFx} + \lambda^TBu + \lambda^TGw - \lambda^Tdx/dt]\,dt\right\} \tag{2.40}$$

This minimization problem is solved, following Wonham's (1968) approach.

Let,

$$x(t) = x_1(t) + x_2(t)$$

such that,

$$dx_1/dt = F(t)x_1(t) + G(t)w(t); \qquad x(0) = x_1(0) + x_2(0) \tag{2.41}$$
$$dx_2/dt = F(t)x_2(t) + B(t)u(t); \qquad x_1(0) \sim N(0,P_0)$$
$$x_2(0) = x_0$$

If $\Xi(t)$ is the error covariance matrix;

$$H(\hat{x}) = \rho + vE\left\{\int_0^T \|\hat{x}-x\|^2\,dt\right\} = \rho + vE\left\{\int_0^T Tr[\hat{x}-x][\hat{x}-x]^T\,dt\right\} =$$
$$= \rho + vE\left\{\int_0^T \Xi(t)\,dt\right\} \tag{2.42}$$

Since x_2 is deterministic and known, $\hat{x}_2 = x_2$ and $\hat{x} = \hat{x}_1 + x_2$ is the total estimate. The measurement equations yield;

$$z_1(t) = z(t) - H(t)x_2(t) = H(t)x_1(t) + v(t) \qquad (2.43)$$
$$z_2(t) = x_2(t)$$

The covariance is obtained ;

$$E\{(\hat{x}-x)(\hat{x}-x)^T\} = E\{(\hat{x}_1-x_1)(\hat{x}_1-x_1)^T\} = P(t)$$

and the estimation problem is reduced to minimize $H(\hat{x}_1)$ subject to:

$$dx_1/dt = F(t)x_1(t) + G(t)w(t); \qquad (2.44)$$
$$z_1(t) = H(t)x_1(t) + v(t)$$

The optimal estimator is assumed of the following form:

$$d\hat{x}_1/dt = A_0(t)\hat{x}_1(t) + K(t)z_1(t), \qquad x(0) = 0;$$

the propagation matrix $A_0(t)$ is found to be

$$A_0(t) = F(t) - K(t)H(t) \qquad (2.45)$$

and the filter gain K(t) satisfies the matrix Riccati equation;

$$dP/dt = [F - KH]P + P[F - KH]^T + GQG^T + KRK^T; \; P(0) = P_0 \qquad (2.46)$$

The value of K(t) is obtained by minimizing eq. (2.42) subject to eq.(2.43) using variational methods, to yield;

$$K(t) = P(t)H(t)^T R(t)^{-1}$$
(2.47)

The estimator is then represented by

$$d\hat{x}/dt = \hat{x}_1(t) + x_2(t) = F(t)\hat{x}(t) + K(t)[z(t) - H(t)\hat{x}(t)]; \; \hat{x}(0)=x_0 \qquad (2.48)$$

It was shown above that the optimal estimator is a Kalman-Bucy filter. It is claimed that the optimal controller now results from the minimization of $H(u/\hat{x})$. It may be interpreted as the optimal deterministic control minimizing J(u) with the states been replaced by their optimal estimates ,

$$u^*(\hat{x}) = - A_1(t)\hat{x}(t) \qquad (2.49)$$

where $A_1(t)$ is the optimal deterministic gain and the minimum value of the differential entropy is;

$$H(u^*/\hat{x}) = \lambda + \mu J(u^*(\hat{x})) \qquad (2.50)$$

To prove that $u^*(\hat{x})$ is really the optimal solution of the LQG it suffices to show that the equivocation $H(\hat{x}/u^*) = 0$ in the equation that follows, and therefore the system is neutral, and the separation theorem holds (Saridis 1977).

$$H(u^*) = H(u^*/\hat{x}) + H(\hat{x}) - H(\hat{x}/u^*)$$

From the worst case densities $p(\hat{x})$ and $p(u^*/\hat{x})$ the density $p(\hat{x}/u^*)$ is obtained, using the chain rule of probabilities,

$$p(\hat{x}/u^*) = \exp\{-\lambda-\rho-\mu[I(u^*(\hat{x}))-I(u(x))] - v\int_0^T \|\hat{x}-x\|^2\, dt\} \qquad (2.51)$$

and $\kappa = \lambda + \rho$, the equivocation is

$$H(\hat{x}/u^*) = \kappa + \mu E\{[I(u^*(\hat{x}))-I(u(x))]\} - vE\{\int_0^T \|\hat{x}-x\|^2\, dt\} =$$
$$= \kappa - \mu\int_0^T Tr[M(t)P(t)]\, dt + v\int_0^T TrP(t)\, dt \qquad (2.52)$$

The last expression is deterministic constant and independent of $u^*(x)$ and x. Furthermore, the constants μ and v are arbitrary, so they may be selected to make the equivocation zero regardless of the selection of $u^*(\hat{x})$ and $\hat{x}$.

$$H(\hat{x}/u^*) \equiv 0 \qquad (2.53)$$

This proves that the system is neutral, the separation theorem holds and that $u^*(\hat{x})$ is truly the optimal control.

In the case of nonlinear systems, the quantity $E\{[I(u^*(\hat{x}))-I(u(x))]\}$ is NOT independent of $u^*(\hat{x})$ and therefore $H(\hat{x}/u^*)$ cannot be set to zero for all times $t\varepsilon[0,T]$ and constant μ and v. This verifies Fel'dbaum's dual control.

2.13 UPPER BOUND OF THE EQUIVOCATION $H[\phi/u^*]$

In the Stochastic Optimal Adaptive Control problem, when x is available for measurement, the Equivocation is reduced to $H[\phi/u^*]$. Since the calculation of $H[\varphi/u^*]$ is practically impossible, it is easy to obtain an **Upper Bound**, assuming a Least squares Identification algorithm, and invoking Jaynes' Principle of Maximum Entropy, twice:

$$p(\phi) = -\alpha - \beta\exp[-\|\phi-\varphi\|^2]$$

$$H[\phi] = \alpha + \beta E\{\|\phi-\varphi\|^2\}$$

and

$$p(u/\phi) = -\gamma - \sigma\exp[-V(x_0,t_0,u,\varphi)]$$

$$H(u/\phi) = \gamma + \sigma E\{V(x_0,t_0,u,\phi)\}$$

$$p(\phi/u)=p(u/\phi)p(\phi)/p(u)= \exp\{-k-\sigma\partial V/\partial\varphi(\phi-\varphi)+(\phi-\varphi)^T[½\sigma\partial^2V/\partial\varphi^2+I](\phi-\varphi)+ +o(\phi-\varphi)\} \tag{2.54}$$

to establish an **Upper Bound of the Equivocation of Active Transmission of Information.**

$$\overline{H}[\phi/u] = C+E\{\sigma\partial V/\partial\varphi(\phi-\varphi)(\phi-\varphi)^T[0.5\sigma\partial^2V/\partial\varphi^2+Q(\phi-\varphi)\} > H[\phi/u] \tag{2.55}$$

C denotes a normalization constant and o(·) is the little o notation. Recursive algorithms of the upper bound may be useful for on-line Adaptive control algorithms.

Other Self-Organizing control algorithms described in Saridis (1977), may also be considered as entropy minimization problems, and upper bounds of the Equivocation may be useful in obtaining approximate solutions. They always provide a worst case solution of the optimization problem.

2.13.1 EXAMPLE: THE UPPER BOUND OF EQUIVOCATION

The following is an illustrative example of the use of approximate entropy measures to evaluate optimal and adaptive control solutions. For simplicity of calculations the following system, with only random initial conditions, is considered;

$$dx_1/dt = x_2 \qquad x(0) \sim N(0,10I)$$
$$dx_2/dt = \varphi x_2 + u \qquad \varphi = -1$$
$$z = x_1$$

and the performance cost;

$$J(u) = E\{\int_0^T (2x_1^2 + 2x_2^2 + ½u^2)\, dt\}$$

The problem is to compare the optimal and adaptive controls using approximate entropy measures, thus utilizing the upper bounds of Equivocation to evaluate their performance for various initial conditions.

a. The Optimal Control

Assuming that φ is known, and some fixed initial conditions x, the optimal solution is obtained by minimizing the entropy function;

$$\mathrm{Min}_u H(x,u) = V(x) = p_{11}x_1^2 + 2p_{12}x_1x_2 + p_{22}x_2^2$$

Using classical control solutions, the above yields;

$$p_{11} = 3, \quad p_{12} = 1, \quad p_{22} = 1$$
$$u^*(x) = -2x_1 - 2x_2$$

and

$$V(x) = 3x_1^2 + 2x_1x_2 + x_2^2$$

b. The Adaptive Control.

The Adaptive Control solution, assuming that φ is known, may be obtained by first identifying φ using a Least Squares algorithm, and then substituting its estimated value to the deterministic Optimal control law.

The Least Squares identification algorithm generates φ, at discrete-time instants, sequentially;

$$\phi(k) = \phi(k-1) + P(k)z(k-1)[z(k) - z(k-1)\phi(k-1)]$$
$$P(k) = P(k-1) - P(k-1)^2z(k-1)^2/[P(k-1)^2z(k-1)^2 + 1]$$

which minimizes;

$$I = E\{\|z(k) - z(k-1)\phi\|^2\}$$

The Optimal control law is obtained by minimizing the deterministic Hamilton-Jacobi-Bellman eq. (2.5), using the estimated parameter φ:

$$u^*(x) = -2x_1 - [\phi + \sqrt{\phi^2+8}]x_2$$

The value of $V(x,\phi) = V(x,u^*,\phi)$ for $u^*(x)$ and ϕ, is given by;

$$V(x,\phi) = [9+\phi+\phi^2+(\phi+1)\sqrt{\phi^2+8}]x_1^2/[\phi+1+\sqrt{\phi^2+8}] + 2x_1x_2 +$$
$$+ [4+\tfrac{1}{2}\phi^2+\tfrac{1}{2}\phi\sqrt{\phi^2+8}]x_2^2/[\phi+1+\sqrt{\phi^2+8}]$$

The values of the performance cost $V(x,\phi)$ are sequentially upgraded as new estimates of φ are obtained.

c. The upper Bound of the Equivocation.

The Equivocation of Active Transmission of Information has been defined to be a function of the probability density;

$$p(\phi/u) = \exp\{ - \sigma[V(x,\phi) - V(x,\varphi)] - \|\phi-\varphi\|^2\}$$

and

$$H(\phi/u) = E\{\sigma[V(x,\phi) - V(x,\varphi)] + \|\phi-\varphi\|^2\}$$

It is practically impossible to calculate this function. Instead the Upper Bound of the Equivocation is calculated and used as a measure to compare the Optimal and Adaptive solutions.

$$H_{ub} = E\{|\sigma\partial V/\partial\varphi\cdot D\varphi| + \tfrac{1}{2}[|\sigma\partial^2 V/\partial\varphi^2| + |I|]D\varphi^2\}$$

where the difference between estimate and true value $D\varphi$ is defined as;

$$D\varphi = \phi-\varphi.$$

For $\sigma = 1$, the components of the Upper Bound eq.(2.54) are;

$$\begin{aligned}\partial V/\partial\varphi = {} & [2\phi^3+3\phi^2+8+2\phi(\phi+1)\surd\phi^2+8]/ \\ & /[2\phi^3+2\phi^2+16\phi+(2\phi^2+2\phi+9)\surd\phi^2+8][x_1+\tfrac{1}{2}x_2^2]\end{aligned}$$

and

$$\begin{aligned}\partial^2 V/\partial\varphi^2 = {} & [-2\phi^4+86\phi^3+190\phi^2+206\phi+ \\ & +(52\phi^5+116\phi^4+33\phi^3+32\phi^2+36\phi+96)\surd\phi^2+8]/ \\ & /[8\phi^6+16\phi^5+140\phi^4+228\phi^3+721\phi^2+800\phi+904+ \\ & +(8\phi^5+16\phi^4+108\phi^3+164\phi^2+352\phi+128)\surd\phi^2+8]\cdot \\ & \cdot[x_1^2+\tfrac{1}{2}x_2^2].\end{aligned}$$

The Upper Bound is then evaluated and compared to the actual Equivocation for values of the parameter φ, and different initial conditions x_{10}, x_{20}, in the Table (2.1).

In this Table IC stands for initial conditions x_0, $X^2 = (x_{10}^2+x_{20}^2)$ and

$$DV = V_{ad} - V_{opt}.$$

TABLE 2.1
UPPER BOUND EVALUATION FOR VARIOUS INITIAL CONDITIONS

IN. CON.	OPT.	ADAP.	DV	$\lvert\partial V/\partial\Phi\rvert$	$\lvert\partial^2 V/\partial\Phi^2\rvert$	X^2	$\bar{H}$
Φ=-0.50, DΦ=0.5							
IC 10	600.	610.25	10.25	0.216	0.204	150.	20.6
IC 5	150.	152.50	2.50	0.216	0.204	37.5	5.325
IC 1	6.	6.30	0.30	0.216	0.204	1.5	0.453
Φ=-0.75, DΦ-0.25							
IC 10	600.	607.2	7.2	0.27	0.22	150.	11.22
IC 5	150.	151.95	1.95	0.27	0.22	37.5	2.85
IC 1	6.	6.27	0.27	0.27	0.22	1.5	0.33
Φ=-0.9, DΦ=0.1							
IC 10	600.	602.10	2.20	0.33	0.232	150.0	5.13
IC 5	150.	150.60	0.60	0.33	0.232	37.5	1.29
IC 1	6.	6.04	0.04	0.33	0.232	1.5	0.06

This example demonstrates the ability of obtaining a measure of closeness between the Adaptive and the Optimal solutions. The acceptance of the Adaptive solution as approximately optimal during the transient period depends on the acceptance of the calculated difference DV.

2.14 CONCLUSIONS

The purpose of this Chapter was to demonstrate the reformulation of the optimal, dual, suboptimal, and adaptive estimation and control using entropy as the cost of performing those tasks. The outcome.was interesting since along with the new interpretation of the control actions at and away of the optimal solutions, analytic justification of the claim of the dual control as the globally optimal solution of the stochastic optimal control problem was obtained.

The other interesting result obtained, was the establishment of the formulation of the general control problem away from the equilibrium, represented by the optimal value, consistent with the theory of chaos, by solving the Generalized Hamilton-Jacobi-Bellman equation resulting from the incompressibility condition of the probability density function in time.

The irreversibility of the control process was established by considering the cost for the control actions as entropy. Once the control action is performed and the cost is paid one cannot reverse the process and resume the ignorance of the solution.

The next chapters will deal with various applications of the entropy formulation of systems control.

2.15 REFERENCES

Feld'baum, A.A. (1965), ***Optimal Control Systems,*** Academic Press, New York.

Jaynes, E.T. (1957), "Information Theory and Statistical Mechanics", *Physical Review*, **Vol.4**, pp. 106.

Kalata, P. Premier, R., (1974), "On Minimal Error Entropy Stochastic Approximation" *International Journal of System Science*, **Vol. 5**, No. 9, pp. 985-986.

Kolmogorov, A.N. (1956), "On Some Asymptotic Characteristics of Completely Bounded Metric Systems", *Dokl Akad Nauk*, SSSR, **Vol. 108**, No. 3, pp. 385-389.

Kumar, P.R., Varaiya, P., (1986), ***Stochastic Systems: Estimation, Identification, and Adaptive Control,*** Prentice Hall Inc. Englewood Cliffs, NJ.

McInroy J.E., Saridis G.N.,(1991), "Reliability Based Control and Sensing Design for Intelligent Machines", in ***Reliability Analysis*** ed. J.H. Graham, Elsevier North Holland, N.Y.

Morse, A.S., (1990), "Towards a Unified Theory of Parameter Adaptive Control: Tunability", *IEEE Trans. on Automatic Control*, **Vol. 35**, No. 9, pp. 1002-1012, September.

Morse, A.S., (1992), "Towards a Unified Theory of Parameter Adaptive Control-Part II: Certainty Equivalence and Implicit Tuning", *IEEE Trans. on Automatic Control*, **Vol. 37**, No. 1, pp. 15-29, January.

Saridis, G.N. (1977), ***Self-Organizing Controls of Stochastic Systems***, Marcel Dekker, New York, New York.

Saridis, G.N. (1979), "Toward the Realization of Intelligent Controls", *IEEE Proceedings*, Vol. 67, No. 8.

Saridis, G. N. (1983), "Intelligent Robotic Control", *IEEE Trans. on Automatic Control*, Vol. 28, No. 4, pp. 547-557, April.

Saridis, G. N., (1985), "Intelligent Control: Operating systems in Uncertain Environments", Chapter 7 in ***Uncertainty and Control***, Ed. J. Ackermann, Springer Verlag, Berlin pp.215-233.

Saridis, G.N. (1988), "Entropy Formulation for Optimal and Adaptive Control", *IEEE Transactions on Automatic Control*, **Vol. 33**, No. 8, pp. 713-721, Aug.

Saridis, G.N., (1991), Architectures for Intelligent Machines" *Proceedings of Workshop for Advanced Robotics*, Beijing, P.R.China, Aug. 23-29.

Saridis, G.N. and Lee, C.S.G. (1979), "Approximation of Optimal Control for Trainable Manipulators", *IEEE Trans. on Systems Man and Cybernetics*, **Vol.8**, No. 3, pp. 152-159, March.

Saridis, G.N. and Valavanis, K.P. (1988), "Analytical Design of Intelligent Machines", *AUTOMATICA the IFAC Journal, **24***, No. 2, pp. 123-133, March.

Tsai, Y.A., Casiello, F.A., Loparo, K.A., (1992), "Discrete-time Entropy Formulation of Optimal and Adaptive Control Problems", *IEEE Transactions on Automatic Control*, **Vol. AC-37**, No. 7, pp. 1083-1088, July.

Wang, F., Saridis, G.N. (1990) "A Coordination Theory for Intelligent Machines" *AUTOMATICA the IFAC Journal, **35***, No. 5, pp. 833-844,Sept.

Weidemann, H.L., (1969), "Entropy Analysis of Feedback Control Systems" in ***Advances in Contol Systems***, C. Leondes Ed., Academic Press, New York NY.

Wonham, W. M., (1968), "On the Separation Theorem of Stochastic Control" SIAM Journal on Control.

Valavanis, K.P., Saridis, G.N., (1992), ***Intelligent Robotic System Theory: Design and Applications***, Kluwer Academic Publishers, Boston, MA.

CHAPTER 3.

REVIEW OF INTELLIGENT CONTROL SYSTEMS

3.1. INTRODUCTION

One of the main applications of the entropy formulation of optimal control is the theory of Intelligent machines, which comprises of the combination of various seemingly disjoint disciplines like task planning, decision making, vision and control. There, the need of a common measure of the cost of performance of the various parts of the system, which proved to be entropy. A survey by Antsaklis' Task Force (1994) supports this claim

In the last few years, Intelligent Machines (Fig. 3.1), proposed by Saridis (1983), have reached a point of maturity to be implemented on a robotic test-bed aimed for space assembly and satellite maintenance. They feature an application of the Theory of Hierarchically Intelligent Control, which is based on the **Principle of Increasing Precision with Decreasing Intelligence (IDI)** to form an analytic methodology, using Entropy as a measure of performance. The original architecture represented a three level system, structured ac-cording to the principle, and using an information theoretic approach (Saridis and Valavanis 1988). The three levels, are (Fig.3.2):

- Organization level
- Coordination level and
- Execution level

representing the original architecture of the system, have not been changed, but their internal architectures have been recently modified to incorporate more efficient and effective structures dictated by experience.

3.1.1 DERIVATION OF THE IDI

At this point it is essential to produce a derivation of IDI, based on Entropy , to show that the Principle is founded on analytic concepts, and therefore has a theoretical basis (Saridis 1989).

The **Principle of Increasing Precision with Decreasing Intelligence (IDI),** is expressed probabilistically by:

$$\text{Prob}(MI,DB) = \text{Prob}(R)$$

where MI is Machine Intelligence, DB is a Data Base, and R is the Flow of Knowledge in an Intelligent Machine.

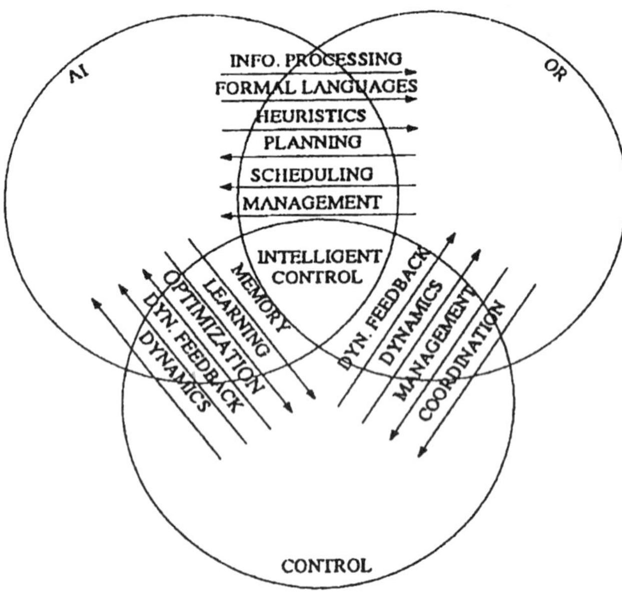

Fig. 3.1 Definition of the Intelligent Control Discipline

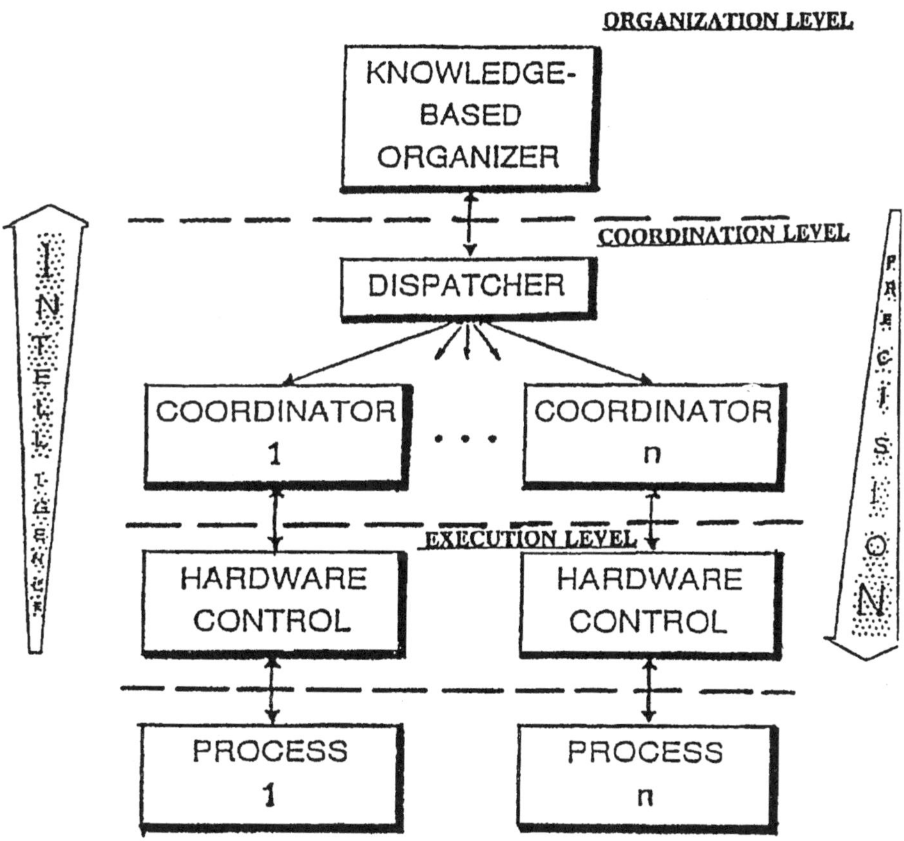

Fig. 3.2 A Hierarchically Intelligent Control System

$$P(MI/DB) \times P(DB) = P(R)$$
$$\ln P(MI/DB) + \ln P(DB) = \ln P(R)$$

Taking the expected values on both sides:

$$H(MI/DB) + H(DB) = H(R) \tag{3.1}$$

where $H(\cdot)$ is the entropy associated with $(\cdot)$.

If MI is independent of the data base DB, then:

$$H(MI) + H(DB) = H(R)$$

which is a manifestation of the **Principle of IDI**.

In the case that P(MI) and P(DB) satisfy Jaynes' Principle of Maximum Entropy, along with P(R):

$$P(R) = \exp(-\alpha_1 - \mu_1 R)$$
$$P(MI/DB) = \exp(-\alpha_2 - \mu_2 MI_{DB})$$
$$P(DB) = \exp(-\alpha_3 - \mu_3 DB)$$

the entropy equation (3.1) is rewritten as:

$$\alpha + \beta MI_{DB} + \mu DB = R \tag{3.1a}$$

where $\alpha = (\alpha_1 - \alpha_2 - \alpha_3)/\mu_1$, $\beta = \mu_2/\mu_1$, and $\mu = \mu_3/\mu_1$.

Equations (3.1) and (3.1a) provide an analytic justification of the increase of Intelligence with the decrease of precision, when R is kept constant. A generalization of that when R is variable is obvious.

3.2 THE ORGANIZATION LEVEL

3.2.1 The Architecture

A Boltzmann machine type neural net, originally proposed for text generation, has been used for the structure that implements the Organization level of an Intelligent Machine developed by Moed and Saridis, (1990). This machine would connect a finite number of letters (nodes) into grammatically correct words (rules), by minimizing at the first layer the total entropy of connections. Replacing the letters at the nodes with words, at second layer, sentences are created. At the third level the words are replaced by sentences at the nodes and so on and so forth until a meaningful text is created.

The functions of the Organizer, following the model of a knowledge based system, comprise of representation, abstract task planning (with minimal knowledge of the current environment), decision making, and learning from experience. All those functions can be generated by a Boltzmann machine similar to the text generating machine, by considering a finite number of primitive elements at the nodes, constituting the basic actions and actors at the representation phase. Strings of these primitives are generated by the Boltzmann machine at the planning phase with the total entropy representing the cost of connections. The selection of the string with minimum entropy is the decision making process, and the upgrading of the parameters of the system by rewarding the successful outcomes through feedback, is the learning procedure. The next to minimum entropy string may be retained as an alternate plan in case of failure of the original or errors created by the environment.

This bottom-up approach, characteristic of natural languages, is extremely simple and effective, utilizing intelligence to replace the complexity of the top-down type task decompositions. The tasks thus generated, are practically independent of the current environment. Information about the present world should be gathered at the Coordination level. An appropriate world model is constructed from sensory and motion information available at that level. However, there the structure of the Dispatcher, designed to interpret the Organizer's strings, monitor and traffic commands among the other Coordinators is highly dependent on the strings which represent the planned tasks.

3.2.2 The Analytic Model

To specify analytically the model of the organizer, it is essential to derive the domain of the operation of the machine for a particular class of problems as in Saridis and Valavanis (1988). Assuming that the environment is known, one may define the following functions on the organization level:

a. Machine Representation and Abstract Reasoning, (RR) is the association of the compiled command to a number of activities and/or rules. A probability function is assigned to each activity and/or rule and the Entropy associated with it is calculated. When rules are included one has active reasoning (inference engine).

In order to generate the required analytic model of this function the following sets are defined:

The set of *commands* $C = \{c_1, c_2, \ldots, c_q\}$ in natural language, is received by the machine as inputs. Each command is compiled to yield an equivalent machine code explained in the next section.

The *task domain* of the machine contains a number n of independent objects.

The set $E = \{e_1, e_2, \ldots, e_m\}$ are *individual primitive events* stored in the long-term memory and representing primitive tasks to be executed. The task domain indicates the capabilities of the machine.

The set $A = \{a_1, a_2, \ldots, a_l\}$ are *individual abstract actions* associating the above events to create sentences by concatenation. They are also stored in the long-term memory.

The set $S = \{s_1, s_2, \ldots, s_n\} = E \cup A$, $n=m+l$, is the group of ***total*** objects which combined, define actions represent complex tasks. They represent the nodes of a Neural net.

A set of *random variables* $X = \{x_1, \ldots, x_n\}$ representing the state of events is associated with each individual object s_i. If the random variable x_i is binary (either 0 or 1), it indicates whether an object s_i is *inactive* or *active*, in a particular activity and for a particular command. If the random variables x_i are continuous (or discrete but not binary) over [0,1], they reflect a membership function in a fuzzy decision making problem. In this work, the x_i's are considered to be binary.

A set of *probabilities* P associated with the random variables X is defined as follows:

$$P = \{P_i = \text{Prob}[x_i = 1];\ I=1,\ldots n\}$$

The probabilities P are known at the beginning of the representation stage. In order to reduce the dimensionality problem of a subset of objects is defined for a given command c_k:

$$S_k = \{s_i;\ P_i \geq a:\ I=1\ldots n\} \subset S \tag{3.2}$$

b. Machine Planning,(P), is ordering of the activities. The ordering is obtained by properly concatenating the appropriate abstract primitive objects $s_i \in S_k$ for the particular command c_k, in order to form the right abstract activities (sentences or text).

The ordering is generated by a Boltzmann machine (Moed Saridis 1990), which measures the average flow of knowledge from node j to node I on the Neural-net by

$$R_{ij} = -\alpha_{ij} - ½E\{w_{ij}x_ix_j\} = -\alpha_{ij} - ½w_{ij}P_iP_j \geq 0 \tag{3.3}$$

The probability due to the uncertainty of knowledge flow into node I, is calculated as in Saridis (1985):

$$p(R_i) = \exp(-\alpha_i - ½\textstyle\sum_j w_{ij}P_iP_j) \tag{3.4}$$

where

$w_{ij} \geq 0$ is the interconnection weight between nodes I and j

$w_{ij} = 0$
$\alpha_i > 0$ is a probability normalizing factor.

The average Flow of Knowledge Ri into node I, is:

$$R_i = \alpha_i + ½E\{ \sum_j(w_{ij}x_ix_j)\} = \alpha_i + ½\sum_j(w_{ij}P_iP_j)$$

with probability $P(R_i)$, (Jaynes' Principle):

$$P(R_i) = \exp[-\alpha_i - ½\sum_j(w_{ij}P_iP_j)]$$

The Entropy of Knowledge Flow in the machine is

$$H(R) = - \sum_i [P(R_i) \ln[P(R_i)] = \sum_i[\alpha_i + ½\sum_j(w_{ij}P_iP_j) \exp[-\alpha_i - ½\sum_j(w_{ij}P_iP_j)] \quad (3.5)$$

The normalizing factor α_i is such that $½^n \le P(R_i) \le 1$.

The entropy is maximum when the associated probabilities are equal, $P(R_ii) = ½^n$ with n the number of nodes of the network. By bounding $P(R_i)$ from below by $½^n$ one may obtain a unique minimization of the entropy corresponding to the most like sequence of events to be selected.

Unlike the regular Boltzmann machines, this formulation does not remove α_i when $P_i = 0$. Instead, the machine operates from a base entropy level $\alpha_i\exp(-\alpha_i)$ defined as the *Threshold Node Entropy* which it tries to reduce (Saridis, Moed 1988).

c. Machine Decision Making,(DM) is the function of selecting the sequence with the largest probability of success.

This is accomplished through a search to connect a node ahead that will minimize the Entropy of Knowledge Flow at that node:

$$H(R_i) = (\alpha_i + ½\sum_j w_{ij}P_iP_j) \exp[-\alpha_i -½\sum_j w_{ij}P_iP_j]$$

A modified genetic algorithm, involving a global random search, has been proposed by Moed and Saridis (1990), as a means of generating the best sequence of events that minimized the uncertainty of connections of the network expressed by the entropy. This algorithm, proven to converge globally compared favorably with other algorithms like the Simulated Annealing and the Random Search.

d. Machine Learning,(ML) (Feedback). Machine Learning is obtained by feedback devices that upgrade the probabilities P_i and the weights w_{ij} by evaluating the performance of the lower levels after a successful iteration.

For y_k representing either P_{ij} or w_{ij}, corresponding to the command c_k, the upgrading algorithms are:

$$y_k(t_k+1) = y_k(t_k) + \beta_k(t_k+1)[\Gamma(t_k+1) - y_k(t_k)] \tag{3.6}$$

$$J_k(t_k+1) = J_k(t_k) + \sigma_k(t_k+1)[V^k{}_{obs}(t_k+1) - J_k(t_k)]$$

where $J_k(t_k)$ is the performance estimate, $V^k{}_{obs}$ is the observed value and

$$P_i : \Gamma_k(t_k+1) = x(t_k)$$

$$w_{ij} : \Gamma_k(t_k+1) = \begin{cases} 1 & \text{if } J = \min\limits_{e} J_e \\ 0 & \text{otherwise} \end{cases} \tag{3.7}$$

e. Memory Exchange (ME), is the retrieval and storage of information from the *long-term memory*, based on selected feedback data from the lower levels after the completion of the complex task.

The above functions may be implemented by a two level Neural net, of which the nodes of the upper level represent the primitive objects s_i and the lower level of primitive actions relating the objects of a certain task. The purpose of the organizer may be realized by a search in the Neural net to connect objects and actions in the most likely sequence for an executable task.

Since it was agreed to use Petri Net Transducers (PNT) to model the coordinators at the next level, a Petri Net generator is required to create the Dispatcher's PNT for every task planned. This can be accomplished by another Boltzmann machine or a part of the existing plan generating architecture.

A graph of the Boltzmann machine with the appropriate symbols is given in Figure 3.3.

3.3. THE COORDINATION LEVEL

3.3.1 The Architecture

The *Coordination level* is a tree structure of *Petri Net Transducers* as coordinators, proposed by Wang and Saridis (1990) with the Dispatcher as the root. Figure 3.4 depicts such a structure. The Petri Net Transducer for the Dispatcher is generated by the Organizer for every specific plan and is transmitted, asynchronously, to the Coordination level along with the plan to be executed. The function of the Dispatcher is to interpret the plan and assign individual tasks to the other coordinators, monitor their operation, and

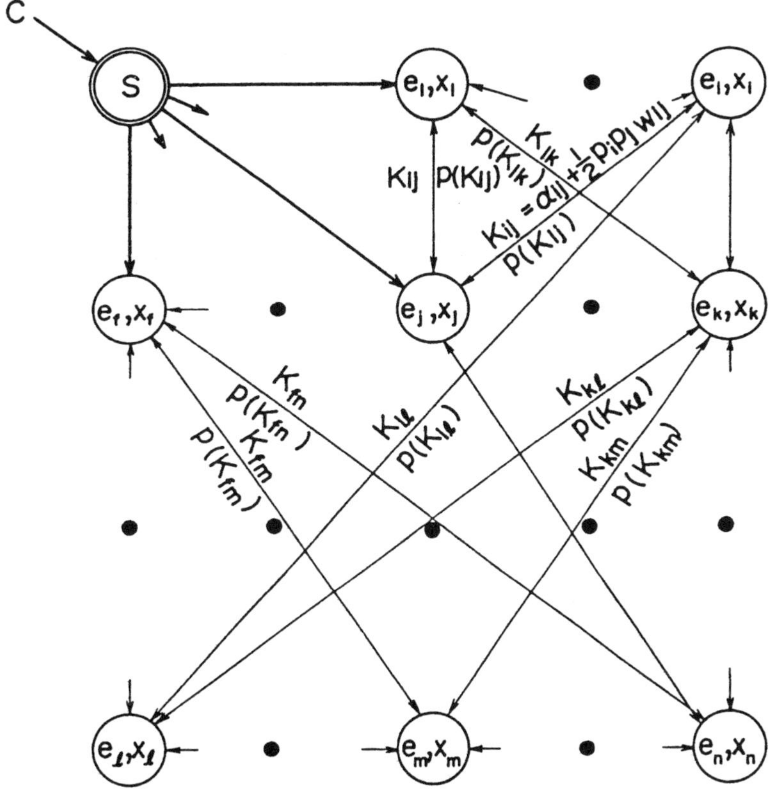

e_i = primitive event

x_i = state of event e_i, $c\{1,0\}$; with prob. p_i

K_i = energy at node i, $= \alpha_i + \frac{1}{2}\sum_j p_i p_j w_{ij}$

w_{ij} = learned weights

$p(K_{ij})$ = probability of connection i-j

Fig. 3.3 The Boltzmann Machine for the Organization Level

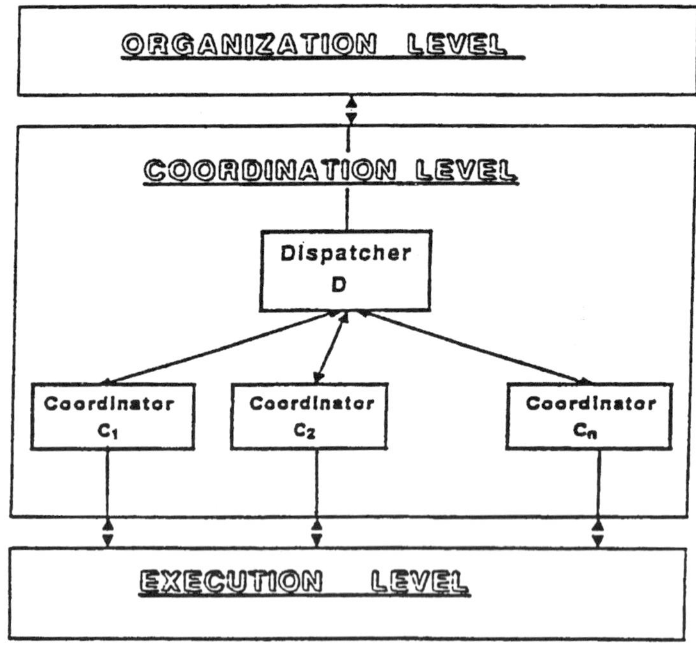

Fig. 3.4 Block Diagram of the Coordination Level

transmit messages and commands from one coordinator to another as needed. As an example, a command is sent to the vision and sensing coordinator to generate a model of the environment, the coordinates of the objects for manipulation to be tabulated, and then transmitted to the motion coordinator for navigation and motion control. This command is executed by having each transition of the associated Petri Nets to initialize a package corresponding to a specific action (Peterson 1977). These packages are stored in short memories associated with each of the coordinators.

The rest of the coordinators have a fixed structure with alternate menus available at request. They communicate commands and messages with each other, through the Dispatcher. They also provide information about reception of a message, data memory location, and job completion.
No data is communicated at the Coordination level, since the task planning and monitoring may be located in a remote station, and such an exchange may cause a channel congestion. A preferred configuration for such situations is that the coordinators with a local dispatcher may be located with the hardware at the work site, while a remote dispatcher, connected to the organizer, interacts with local one from a remote position. Figure 3.5 depicts this architecture. This concept simplifies the communication problem considerably, since only short messages are transmitted back and forth through a major channel between local and remote stations, requiring a narrow bandwidth. An example of the effectiveness of such an architecture may be demonstrated in space construction, where robots work in space while task planning and monitoring is done on earth.

Even though, there is no limitation to the number of coordinators attached to the Dispatcher, only the following ones are planned for an Intelligent Robot for space applications.

Vision and Sensory Coordinator. This device coordinates all the sensory activities of the robot, with cameras and lazers, and generates information of the world model in Cartesian coordinates.

Motion Control Coordinator. This device receives control, object and obstacle information and uses it to navigate and move multiple robotic arms and other devices, for object manipulation and task execution. It also assigns the appropriate operations on the data acquired for the desired application.

Planning Coordinator. The task plans, optimal and alternate gene-rated by the Organizer are stored in this device for proper monitoring of execution and possible error recovery in cases of failure of the system.

Grasping Coordinator. This device coordinates the grippers of the arms and interfaces the proximity sensors for effective grasping.

Entropy measures, are developed by McInroy and Saridis (1991), at each coordinator such that they may be used to minimize the complexity and improve the reliability of the system. A typical PNT system for the Coordination level of an Intelligent Robot as proposed by Wang and Saridis (1990), is given in Figure 3.5.

3.3.2 The Analytic Model

Petri nets have been proposed as devices to communicate and control complex heterogenous processes. These nets provide a communication protocol among stations of the process as well as the control sequence for each one of them. Abstract task plans, suitable for many environments are generated at the organization level by a grammar created by Wang and Saridis (1990):

$$G = (N, \textstyle\sum_o, P, S) \tag{3.8}$$

where

$N = \{S, M, Q, H\}$ = Non-terminal symbols
$\sum_o = \{A_1, A_2, \ldots A_n\}$ = Terminal Symbols (activities)
P = Production rules

Petri Net Transducers (PNT) proposed first by Wang and Saridis (1990), are Petri net realizations of the *Linguistic Decision Schemata* introduced by Saridis and Graham (1984), as linguistic decision making and sequencing devices. They were realized on a robotic system by Wang et al (1990) They are defined as 6-tuples:

$$M = (N, \textstyle\sum, \delta, G, \mu, F) \tag{3.9}$$

where

$N = (P, T, I, O)$ = A Petri net with initial marking μ,
$\sum$ = a finite input alphabet
δ = a finite output alphabet
σ = a translation mapping from $T \times (\sum \cup \{\backslash\})$ to finite sets of δ^* and $F \subset R(\mu)$ a set of final markings.

A *Petri Net Transducer (PNT)* is depicted in Figure 3.6. Its input and output languages are *Petri Net Languages (PNL)*. In addition to its on-line decision making capability PNT's have the potential of generating communication protocols, learning by feedback, ideal for the communication and control of coordinators and their dispatcher in real time. Their architecture is given in Figure 3.7, and may follow a scenario suitable for the implementation of an autonomous intelligent robot.

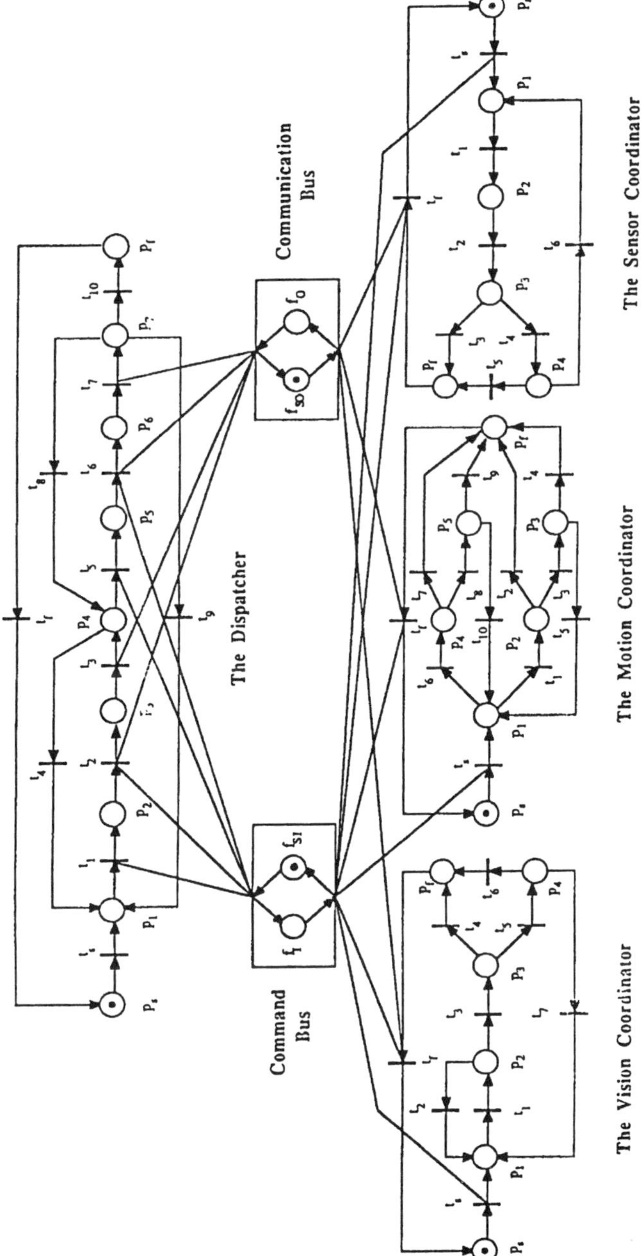

Fig. 3.5 Petri Net Diagram of a Typical Coordination Level

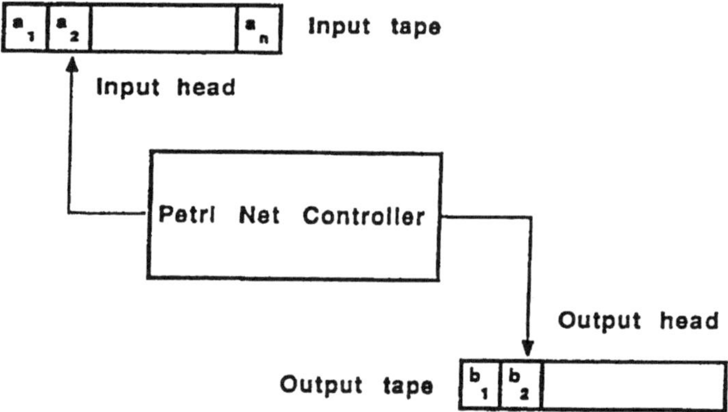

Fig. 3.6 A Petri Net Transducer

Figure 3.8 depicts the Petri Net Structure of a typical *Coordination Structure (CS)* of an intelligent robot. This structure is a 7-tuple:

$$CS = (D, C, F, R_D, S_D, R_C, S_C) \tag{3.10}$$

where

$D = (N_d, \sum_o, \delta_o, G_d, \mu_d, F_d)$ = The PNT dispatcher
$C = \{C_1,...C_n\}$ = The set of coordinators
$C_i = (N^i_c, \sum^i_c, \delta^i_c, G^i_c, F^i_c)$ = the ith PNT coordinator
$F = U^n_{i=1}\{f^i_I, f^i_{SI}, f^i_O, f^i_{SO}\}$ = A set of connection points
R_D, R_C = Receiving maps for dispatcher and coordinators
S_D, S_C = Sending maps for dispatcher and coordinators

Decision making in the coordination structure is accomplished by *Task Scheduling* and *Task Translation*, e.g., for a given task find σ an enabled t such that $\sigma(t,a)$, is defined and then select the right translation string from $\sigma(t,a)$ for the transition t.
The sequence of events transmitted from the organization level is received by the dispatcher which requests a world model with coordinates from a vision coordinator.

The vision coordinator generates appropriate database and upon the dispatcher's command communicates it to the planning coordinator which set a path for the arm manipulator. A new command from the dispatcher sends path information to the motion controller in terms of end points, constraint surface and performance criteria. It also initializes the force sensor and proximity sensor control for grasp activities. The vision coordinator is then switched to a monitoring mode for navigation control.

The PNT can be evaluated in *real-time* by testing the computational complexity of their operation which may be expressed uniformly in terms of entropy. Feedback information is communicated to the coordination level from the execution level during the execution of the applied command. Each coordinator, when accessed, issues a number of commands to its associated execution devices (at the execution level). Upon completion of the issued commands feedback information is received by the coordinator and is stored in the *short-term memory* of the coordination level.

This information is stored in the short-term memory of the coordination level. This information is used by other coordinators if necessary, and also to calculate the individual, accrued and overall accrued costs related to the coordination level. Therefore, the feedback information from the execution to the coordination level will be called *on-line*, *real-time* feedback information.

The performance estimate and the associated subjective probabilities are updated after

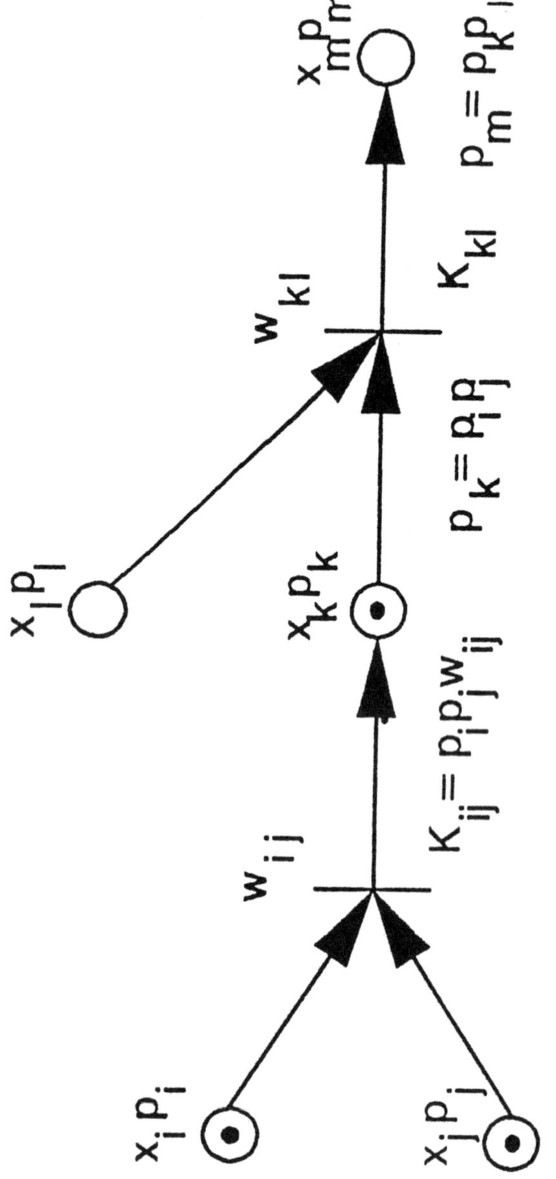

Fig. 3.7 Entropy Measures of a Petri Net Transducer: $H = \sum_{ij} K_{ij}$

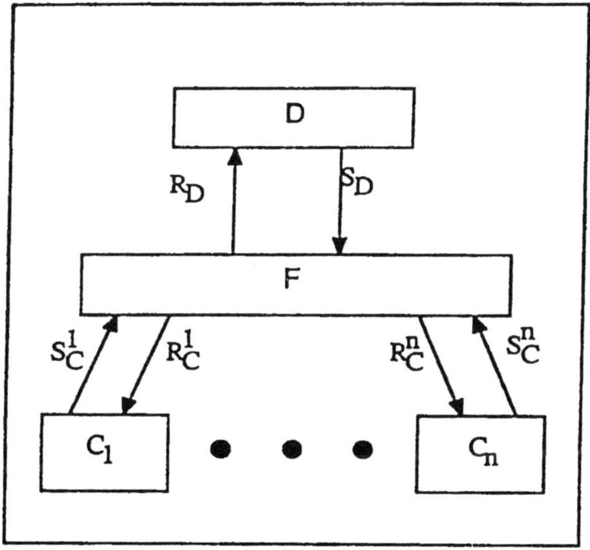

Fig. 3.8 The Coordination Structure

the k_{ij}-th execution of a task $[(u_t,x_t)_i,S_j]$ and the measurement of the estimate of the observed cost J_{ij}:

$$J_{ij}(k_{ij}+1) = J_{ij}(k_{ij})+\beta(k_{ij}+1)[J_{obs}(k_{ij}+1)-J_{ij}(k_{ij})] \tag{3.11}$$

$$P_{ij}(k_{ij}+1) = P_{ij}(k_{ij})+\mu(k_{ij}+1)[\Gamma_{ij}(k_{ij}+1)-P_{ij}(k_{ij})]$$

where

$$\Gamma_{ij} = \begin{cases} 1 & \text{if } J_{ij} = \text{Min} \\ 0 & \text{elsewhere} \end{cases}$$

and β and μ are harmonic sequences. Convergence of this algorithm is proven in Saridis and Graham (1984).

The *learning process* is measured by the entropy associated to the subjective probabilities.

If

$$H(M) = H(E) + H(T/E) \tag{3.12}$$

where H(E) is the environmental uncertainty and H(T/E) is the pure translation uncertainty. Only the last term can be reduced by learning. Figures 3.7 and 3.9 show the generation of such entropies

3.4. THE EXECUTION LEVEL.

3.4.1 The System and the Architecture

The *Execution level* contains all the hardware required by the Intelligent Machine to execute a task. There is a one-to-one correspondence between hardware groups and coordinators. Therefore their structure is usually fixed. This level also contains all the drivers, VME buses, short memory units, processors, actuators and special purpose devices needed for the execution of a task. After the successful completion of a job feedback information is gene-rated at this level for evaluation and parameter updating of the whole machine. Complexity dominates the performance of this level. Since *precision* is proportional to *complexity*, it also defines the amount of effort required to execute a task. It has been shown that all the activities of this level can be measured by entropy, which may serve as a measure of complexity as well. Minimization of local complexity through feedback, may serve as local design procedure.

The localization of data exchange at this level provides a means of efficient remote control of the Intelligent Machine, (Figure 3.9).

The following hardware groups are available:

<u>The Vision and Sensory System.</u> Such a system may consist of two fixed cameras, two penlight cameras on the wrist of one PUMA arm, and a lazer rangefinder. They are all controlled by a Datacube with a versatile menu of various hardwired functions and a VME bus for internal communications. The functions assigned to them, e.g. create a world model in cartesian space, find the fiducial marks on the object to be manipulated, or track a moving object are supported by software specialized for the hardware of the system. Calibration and control of the hardware is an important part of the system. Since we are dealing with information processing the system's performance can be easily measured with entropy. Actual data for visual servoing can be generated on the VME bus and transmitted through the Dispatcher to the Motion Control system. Direct connection of the VME bus with the Motion Control System is planned in the future.

<u>The Motion Control System.</u> This system is a unified structure for cooperative motion and force control for multiple arm manipulation. Since motion affects force but not vice versa, motion control is designed independent of the constraint forces, and force control by treating inertial forces as disturbance. Integral force feedback is used with full dynamics control algorithms. The resulting system, hierarchically integrates the execution algorithms in planning, interaction, and servo control. It works together with the VXWORKS software and provides most of the transformations, and other kinematics and dynamics tools needed for servoing and manipulation. In earlier work it was shown that the control activities can be measured by entropy (1983). Therefore the measure of performance of the Motion Control System is consistent with the rest of the architecture of the Intelligent Machine.

<u>The Grasping System.</u> This system is planned to be separate from the Motion Control System. It would involve the grasping operations, the information gathering from various proximity sensors, and integration of these activities with the gripper motion control. It will be driven by a special coordinator, and provide information back of proper grasping for job control purposes.

3.4.2 Entropy Formulation of Motion Control.

The cost of control at the hardware level can be expressed as an entropy which measures the uncertainty of selecting an appropriate control to execute a task. By selecting an optimal control, one minimizes the entropy, e.g., the uncertainty of execution. The entropy may be viewed in the respect as an energy in the original sense of Boltzmann, as in Saridis (1988).

Optimal control theory utilizes a non-negative functional of the state of the system $x(t) \varepsilon \Omega_x$ the state space, and a specific control $u(x,t) \varepsilon \Omega_u \times T$; $\Omega_u \subset \Omega_x$ the set of all admissible feedback controls, to define the performance measure for some initial conditions $x_0(t_0)$,

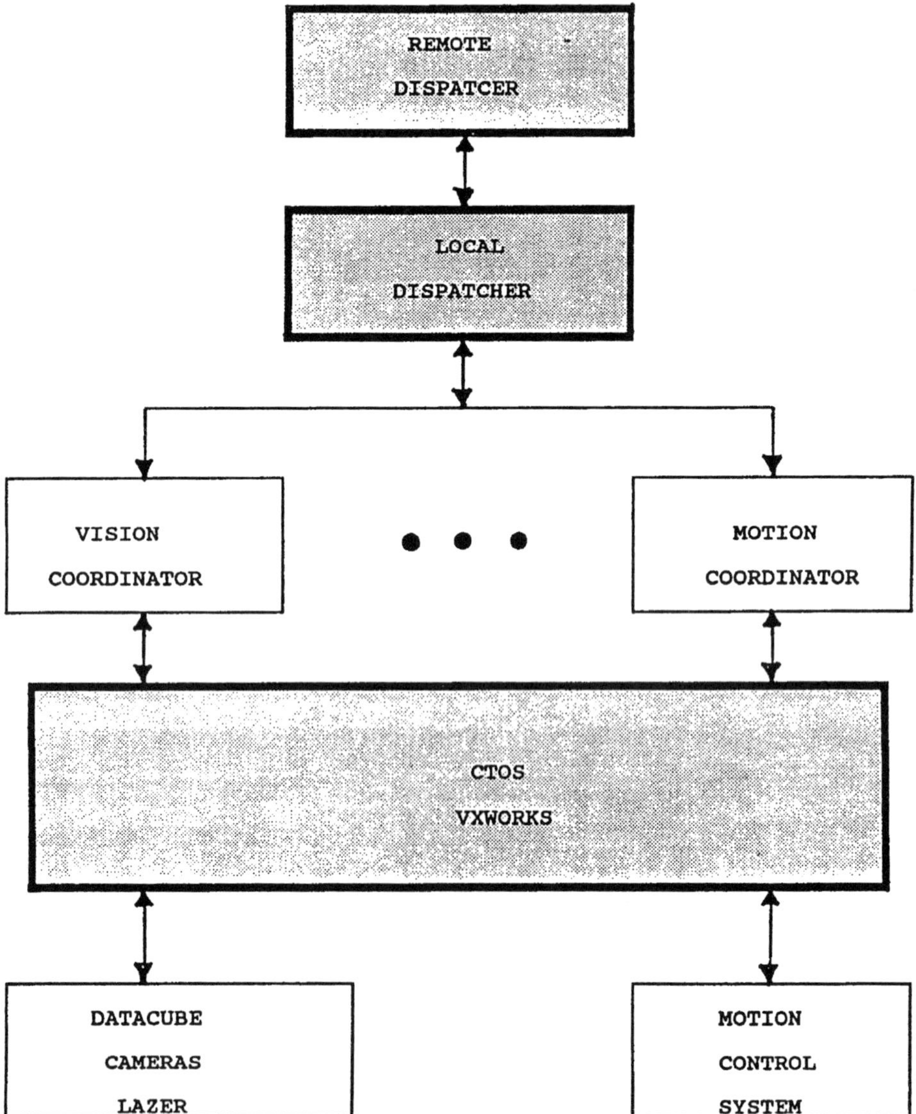

Fig. 3.9 Telerobotics Tested Configuration

representing a generalized energy function, of the form:

$$V(x_0,t_0) = E\{\int_{t0} L(x,t;u(x,t))\,dt\} \tag{3.13}$$

where $L(x,t;u(x,t)) > 0$, subject to the differential constraints dictated by the underlying process

$$\begin{aligned} dx/dt &= f(x,u(x,t),w,t); \quad x(t_0) = x_0 \\ z &= g(x,v,t); \qquad x(t_f) \in M_f \end{aligned} \tag{3.14}$$

where x_0, $w(t)$, $v(t)$ are random variables with associated probability densities $p(x_0)$, $p(w(t))$, $p(v(t))$ and Mf a manifold in Ω_x. The trajectories of the system (4.2) are defined for a fixed but arbitrarily selected control $u(x,t)$ from the set of admissible feedback controls Ω_u.

In order to express the control problem in terms of an entropy function, one may assume that the performance measure $V(x_0,t_0,u(x,t))$ is distributed in u according to the probability density $p(u(x,t))$ of the controls $u(x,t) \in \Omega_u$. *The differential entropy* $H(u)$ corresponding to the density is defined as

$$H(u) = -\int_{\Omega u} p(u(x,t)) \ln p(u(x,t))\,dx$$

and represents the uncertainty of selecting a control $u(x,t)$ from all possible admissible feedback controls Ω_u. The optimal performance should correspond to the maximum value of the associated density $p(u(x,t))$. Equivalently, the optimal control $u^*(x,t)$ should minimize the entropy function $H(u)$.

This is satisfied if the density function is selected to satisfy *Jaynes' Principle of Maximum Entropy* (1957), e.g.,

$$p(u(x,t)) = \exp\{-\lambda - \mu V(x_0,t_0;u(x,t))\} \tag{3.15}$$

where λ and μ are normalizing constants.

It was shown by Saridis (1988), that the expression $H(u)$ representing the entropy for a particular control action $u(x,t)$ is given by:

$$H(u) = \int_{\Omega u} p(x,t;u(x,t))V(x_0,t_0;u(x,t))\,dx = \lambda + \mu V(x_0,t_0;u(x,t)) \tag{3.16}$$

This implies that the average performance measure of a feedback control problem corresponding to a specifically selected control, is an entropy function. The optimal control $u^*(x,t)$ that minimizes $V(x_0,t_0;u(x,t))$, maximizes $p(x,t;u(x,t))$, and consequently minimizes the entropy $H(u)$.

$$u^*(x,t) : E\{V(x_0,t_0;u^*(x,t))\} = \min \int_{\Omega u} V(x_0,t_0;u(x,t))p(u(x,t))\,dx \tag{3.17}$$

This statement is the generalization of a theorem proven by Saridis [16], and establishes equivalent measures between information theoretic and optimal control problem and provides the information and feedback control theories with a common measure of performance.

3.4.3 Entropy Measure of the Vision System

The optimal control theory designed mainly for motion control, can be implemented for vision control, path planning and other sensory system pertinent to an Intelligent Machine by slightly modifying the system equations and cost functions. After all one is dealing with real-time dynamic systems which may be modeled by a dynamic set of equations.

A *Stereo Vision* system of a pair of cameras mounted at the end of a robot arm, may be positioned at i=1,..N different view points to reduce problems with noise, considered one at a time due to time limitations. The accuracy of measuring the object's position depends upon its relative position in the camera frame. Consequently, each viewpoint will have different measurement error and time statistics. These statistics may be generated to define the uncertainty of the measurement of the Vision system as in McInroy and Saridis (1991).

For a point c of the object, the measurement error of its 3-D position in the camera coordinate frame e_{pc} is given by:

$$e_{pc} = M_c\, n_c \tag{3.18}$$

where n_c is the 3-D image position errors, and M_c an appropriate 3X3 matrix, depending on the position of the object.

The linearized orientation error is given by:

$$\delta = (M^T M)^{-1} M^T M' F n \tag{3.19}$$

where

δ is the orientation error in the camera frame,
M is a matrix formed from camera coordinate frame positions,
M' is a constant matrix,
F is the matrix formed from the camera pameters and measured positions,
n is the vector of the image position errors at the four points.

A vector containing the position and orientation errors due to image noise is given by:

$$e_c = [e^T_{pc}\delta^T]^T = Ln \tag{3.20}$$

where L depends on the camera parameters and the four measured camera frame positions of the points. The statistics of the image noise n, due to individual pixel errors are assumed to be uniformly distributed. Assuming that feature matching centroids is used by the vision system, its distributions tend to be independent Gaussian, due to the Central Limit Theorem.

$$n \approx N(0,C_v) \text{ and } e_c \approx N(0,LC_vL^T) \tag{3.21}$$

The time which each vision algorithm consumes is also random due to the matching period. Therefore the total vision time, for the ith algorithm that includes camera positioning time, image processing time, and transformation to the base frame, is assumed Gaussian:

$$t_{vi} \approx N(\mu_{tvi},\sigma^2_{tvi}). \tag{3.22}$$

Once the probability density functions are obtained, the resulting Entropies $H(t_{vi})$, and $H(e_c)$, are obtained in a straight forward manner for the ith Algorithm (McInroy and Saridis 1991):

$$\begin{aligned} H(t_{vi}) &= \ln\sqrt{2\pi e\sigma^2_{tvi}}) \\ H(e_c) &= \ln\sqrt{(2\pi e)^6}\det[C_v] + E\{\ln[\det L_i]\} \end{aligned} \tag{3.23}$$

The total Entropy, may be used as a measure of uncertainty of the Vision system (imprecision), and can be minimized with respect to the available system parameters:

$$H(V) = H(t_{vi}) + H(e_c). \tag{3.24}$$

3.5 CONCLUSIONS

The architecture described in this chapter is a natural evolution of the architecture originally proposed by Saridis (1979), using entropy as the cost of performance. It was debated that this architecture does not differ philosophically from the architectures proposed by Albus, Brooks, Meystel and others as reported in the Report of the Task Force on Intelligent Controls (1989). Albus' algorithm (1975) utilizes several levels of the same architecture to structure a brain type software, which is hierarchical in time and detail (precision) of execution. This represents a much more elaborate structure than the Saridis' model. Meystel (1986) has created an AI type of model where the more detailed activities are nested within the less detailed ones (precision). The difference with the Saridis' model is not its complexity, but that there is no analytic model associated. Finally Brooks' model is purely heuristic, based on training by learning. However, if one looks more carefully, one may see that they all follow the same philosophy and only the respective approach differs.

The model presented in this chapter, based on analytic methods, and therefore appealing to the engineer, uses optimization methods to obtain execution of requested tasks with adaptable interaction with a human. Its main contribution though is that this system is been successfully implemented and that the resulting structure is extremely efficient, effective, versatile, capable for remote operation as compared to other proposed architectures.

3.6 REFERENCES

Albus, J.S. (1975), "A New Approach to Manipulation Control: The Cerebellar Model Articulation Controller", *Transactions of ASME, J. Dynamics Systems, Measurement and Control,* ***97***, pp. 220-227.

Antsaklis P. Chair (1994), "Defining Intelligent Control" Report of the Task Force on Intelligent Control, ***IEEE Control Systems Magazine*** Vol. 14, No. 3, p. 4.

Jaynes, E.T. (1957), "Information Theory and Statistical Mechanics", *Physical Review*, pp. 106, 4.

McInroy J.E., Saridis G.N.,(1991), "Reliability Based Control and Sensing Design for Intelligent Machines", in ***Reliability Analysis*** ed. J.H. Graham, Elsevier North Holland, N.Y.

Meystel, A. (1986), "Cognitive Controller for Autonomous Systems", *IEEE Workshop on Intelligent Control 1985*, p. 222, RPI, Troy, New York.

Moed, M.C. and Saridis, G.N. (1990), "A Boltzmann Machine for the Organization of Intelligent Machines", *IEEE Transactions on Systems Man and Cybernetics,* ***20***, No. 5, Sept.

Peterson, J.L. (1977), "Petri-Nets", *Computing Survey*, 9, No. 3, pp. 223-252, September.

Saridis, G.N. (1977), ***Self-Organizing Controls of Stochastic Systems***, Marcel Dekker, New York, New York.

Saridis, G.N. (1979), "Toward the Realization of Intelligent Controls", *IEEE Proceedings,* ***67***, No. 8.

Saridis, G. N. (1983), "Intelligent Robotic Control", *IEEE Trans. on AC,* ***28***, 4, pp. 547-557, April.

Saridis, G.N. (1985), "Foundations of Intelligent Controls", *Proceedings of IEEE Workshop on Intelligent Controls*, p. 23, RPI, Troy, New York.

Saridis, G.N. (1988), "Entropy Formulation for Optimal and Adaptive Control", *IEEE Transactions on AC, 33*, No. 8, pp. 713-721, Aug.

Saridis, G.N. (1989), "Analytic Formulation of the IDI for Intelligent Machines", *AUTOMATICA the IFAC Journal, 25*, No. 3, pp. 461-467.

Saridis, G.N. and Graham, J.H. (1984), "Linguistic Decision Schemata for Intelligent Robots", *AUTOMATICA the IFAC Journal,* ***20***, No. 1, pp. 121-126, Jan.

Saridis, G.N. and Moed, M.C. (1988), "Analytic Formulation of Intelligent Machines as Neural Nets", *Symposium on Intelligent Control*, Washington, D.C., August.

Saridis, G.N. and Stephanou, H.E. (1977), "A Hierarchical Approach to the Control of a Prosthetic Arm", *IEEE Trans. on SMC,* ***7***, No. 6, pp. 407-420, June.

Saridis, G.N. and Valavanis, K.P. (1988), "Analytical Design of Intelligent Machines", *AUTOMATICA the IFAC Journal, 24*, No. 2, pp. 123-133, March.

Wang, F., Kyriakopoulos, K., Tsolkas T., Saridis, G.N., (1990) "A Petri-Net Coordination Model of Intelligent Mobile robots" *CIRSSE* ***Technical Report #50***, Jan.

Wang, F., Saridis, G.N. (1990) "A Coordination Theory for Intelligent Machines" *AUTOMATICA the IFAC Journal,* ***35,*** No. 5, pp. 833-844, Sept.

CHAPTER 4.

RELIABILITY AS ENTROPY

4.1. SELECTING RELIABLE PLANS:

Reliability theory is an extremely powerful tool developed to assist system designers to generate safe and precise products that can be duplicated , and guaranteed to perform free of malfunctions as possible. Analytic techniques are derived in this chapter which allow reliable plans to be automatically selected. It does not discuss the entire generation of design plans; rather it concentrates on the uncertainty analysis of candidate designs, such that a highly reliable candidate may be identified and used. Entropy is used as a universal measure of system performance. It is based on the concept of uncertainty and serves to uniformly evaluate design procedures. It provides new means of evaluating designs and guarantee effective performance.

Reliability is also a valuable criterion for designing autonomous intelligent control systems, and can be expressed in terms of entropy. For robotic components, such as a particular vision algorithm for pose estimation or a joint controller methods are explained for directly calculating the reliability. However, these methods become excessively complex when several components are used together to complete a plan. Consequently, entropy minimization techniques are used to estimate which complex tasks will perform reliably. In the next chapters we first develop the tools for directly calculating the reliability of subsystems with methods that use entropy minimization. This approach greatly facilitates the analysis of the design procedure which is subsequently explained. Since these subsystems are used together to accomplish complex tasks, we then show how complex tasks can be efficiently evaluated.

The novelty of the presentation of Reliability theory in this chapter is its derivation from the entropy point of view, and therefore its analytic formulation. This approach presents a quantitative methodology which is elegant and flexible and facilities the autonomous selection of execution plans of Intelligent Machines. Reliability theory using entropy measures was first developed in conjunction with the theory of Intelligent Machines (Saridis 1995).

4.2 DEFINITION OF RELIABILITY

Reliability has been used in science and engineering as a criterion of dependence on the quality of a process. A suitable definition is given by Harr (1987):

> **Reliability is the probability of an object (item or system) performing its required function adequately for a specified period of time under stated conditions.**

$$R = \text{Prob}\{g(x)>0\} \tag{4.1}$$

where g(x) is the proposed function of the system and x is its state at time t. Such a definition suggests reliability as a performance criterion if g(x)>0 is considered as the goal to be accomplished by the associated system. It also indicates the probabilistic nature of reliability which immediately suggests its relation to entropy. Both those remarks may associate reliability to the design of Intelligent Control systems which will be discussed in the sequel.

Intelligent Control systems were discussed in the previous chapter under a general unspecified performance criterion. Reliability may be used as the criterion of an acceptable design if eq. (4.1) is satisfied, or it may be used directly as a performance criterion by itself. The interpretation of Reliability as entropy will facilitate the derivation of simple analytic expressions to serve either of the two goals, since they prove to be much simpler than the previously defined analytic techniques.

4.3 RELIABILITY MEASURES

The problem we are dealing with applies to a system described by the state equation

$$dx/dt = f(x,u,t) \tag{4.2}$$

where, as previously stated, x is the n-dim. state vector and u is the m-dim. control input. Then define the error between the actual and the desired state as;

$$\tilde{x} = x_d - x \tag{4.3}$$

In the past several methods have been proposed to evaluate the reliability of a system by calculating the probability that satisfies eq. (4.1). Such methods are described in McInroy, Musto and Saridis (1995) and are summarized here:

Monte Carlo Simulation, assigns reliability as the frequency of successful experiments

$$g(x_i)>0 \qquad I = 1,2...n \tag{4.4}$$

Maximum Likelihood, estimates the parameters of an assumed probability distribution, preferably normal, for eq. (4.1) and assigns reliability its covariance.

Reliability Lower Bounds. This method assumes that the argument of eq. (4.1) is replaced by bounds l_i, forming a hyper rectangle, and reliability becomes:

$$R = \text{Prob}\{\cap_i \|\tilde{x}_i\| \le l_i, I = 1,2...n\} = \prod_1^n \int_{-l_i}^{l_i} p(\tilde{x})\, d\tilde{x}_i \tag{4.5}$$

where the density $p(\tilde{x})$ is not known or difficult to obtain. A more convenient form for calculation purposes is given by an inscribed hyperelipsoid as in Fig. 4.1, defining a lower bound of reliability (McInroy Saridis 1990):

$$R_c = \text{Prob}\{\tilde{x}^T Q \tilde{x} \leq 1\}, \qquad Q = \begin{bmatrix} 1/l_1^2 & & 0 \\ & \cdots & \\ 0 & & 1/l_n^2 \end{bmatrix} \tag{4.6}$$

The following theorem due to McInroy (1990) gives a lower bound on reliability:

Theorem 4.1 Given a random vector $\tilde{x} \sim N(0,C_x) \in \Re^n$, $C\tilde{x}>0$, then a lower bound on the reliability $R=\text{Prob}\{\tilde{x}^T Q \tilde{x} \leq 1\}$, where $Q \geq 0 \in \Re^{n \times n}$, Q symmetric, is given by

$$R \geq R_{lb} = \chi_n^2 \left(1/\max \lambda(C_{\tilde{x}} Q) \right) \tag{4.7}$$

where $\chi_n^2(.)$ denotes the chi-squared distribution function with n degrees of freedom, and max $\lambda(.)$ denotes the maximum eigenvalue .

These reliability methods are not so easy to calculate so that we may use entropy to simplify calculations.

4.4 ENTROPY MEASURES OF RELIABILITY

The entropy associated with a density function p(x) is given by

$$H = -\int_{-\infty}^{\infty} p(x) \ln p(x)\, dx \tag{4.8}$$

The lower bound measure may now be derived in terms of entropy by defining the reliability-based cost function which provides a measure of system performance

$$V = \tilde{x}^T Q \tilde{x} \tag{4.9}$$

then the reliability R_c of the lower bound is

$$R_c = \int_0^1 p(V)\, dV \tag{4.10}$$

In general p(V) the probability density associated with V is unknown. A worst case density can be assigned using a variant of Jaynes Principle of Maximum Entropy (Saridis 1995) which under the worst case assumptions yield the least biased density

$$p(V) = [E\{V\}]^{-1} \exp[-V/E\{V\}] \tag{4.11}$$

Then the worst case entropy is given by

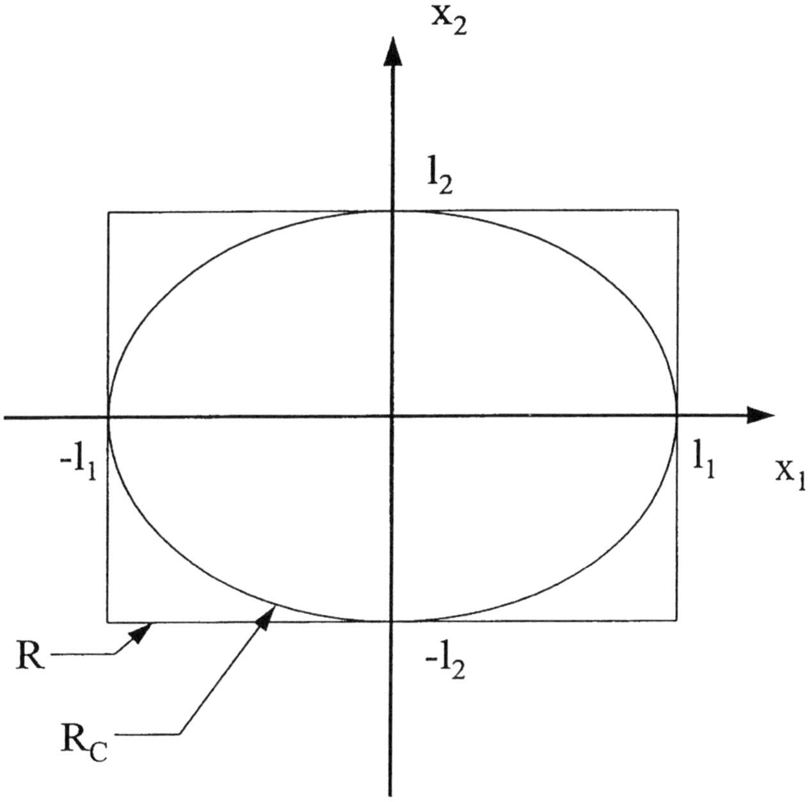

Fig. 4.1 Actual and Lower Bound Reliability

$$H(V) = - \int_{-\infty}^{\infty} p(V)\, \ln p(V)\, dV = \ln E\{V\} + 1 \qquad (4.12)$$

a monotonically increasing function of E{V}. For this density the reliability is

$$R_c = \int_0^1 p(V)dV = \int_0^1 [E\{V\}]^{-1} \exp[-V/E\{V\}]dV = 1 - \exp[-V/E\{V\}] \qquad (4.13)$$

Using Chebychev inequality, reliability is a monotonically decreasing function of E{V}

$$R_c = 1 - E\{V\} \qquad (4.14)$$

If in addition the covariance $C_{\tilde{x}}$ of the error $\tilde{x}$ is known then

$$E\{H(V)\} = E\{\tilde{x}^T Q\, \tilde{x}\} = \mathrm{Trace}\{Q\, C_{\tilde{x}}\} \qquad (4.15)$$

Theorem 4.2 If two alternative systems A and B are compared by their entropies

$$H_B(V) \le H_A(V) \qquad (4.16)$$

then

$$R_{cB} \ge R_{cA} \qquad (4.17)$$

This is true because

$$R_c = 1 - C \exp H(V) \quad C = e^{-1} \qquad (4.18)$$

In conclusion entropy can be used as a measure of lower bound system reliability.

4.5 THE LOOSER LOWER BOUND

Theorem 4.2 provides the relation between the entropy of the cost V and a bound on reliability. However, this entropy involves a weighted combination of the system state information, than the unweighted entropy of its states. If a stricter loose lower bound is used a direct analogy between this bound of reliability and the entropy of the states is obtained.

$$R_{lb} = \mathrm{Prob}\{\lambda_{max}\, \tilde{x}^T \tilde{x} \le 1\} \le \mathrm{Prob}\{\tilde{x}^T Q\, \tilde{x} \le 1\} \qquad (4.19)$$

which follows from the inequality

$$_{max.}\ \tilde{x}^T \tilde{x} \ge \tilde{x}^T Q\, \tilde{x}$$

The loose lower bound of reliability considers the probability of the state variable errors within the maximum radius hyper sphere fully inscribed in the hyper ellipsoid of allowable errors described by the tolerance constraints (see Fig. 4.2).

A new cost function is now defined

$$V_{lb} = {}_{max.}\, \tilde{x}^{T} \tilde{x} \tag{4.20}$$

with a probability density function and entropy respectively

$$p(V_{lb}) = [E\{V_{lb}\}]^{-1} \exp[-V_{lb}/E\{V_{lb}\}] \tag{4.21}$$

$$H(V_{lb}) = -\int_{-\infty}^{\infty} p(V_{lb}) \ln p(V_{lb})\, d\, V_{lb} = \ln E\{V_{lb}\} + 1 \tag{4.22}$$

$$E\{V_{lb}\} = {}_{max.}\, E\{\tilde{x}^{T} \tilde{x}\} = {}_{max.}\, (\mathrm{Trace}\{C_{x}\}) \tag{4.23}$$

Substituting (4.23) into (4.22) one gets the expression for the entropy

$$H(V_{lb}) = \ln({}_{max.}\, (\mathrm{Tr}\{C_{x}\}))+1 = \ln\lambda_{max}+\ln\, (\mathrm{Tr}\{C_{x}\})+1 \tag{4.24}$$

Using the same variant of Jaynes Principle of Maximum Entropy (Saridis 1995) which under the worst case assumptions yield the least biased density

$$H(\tilde{x}^{T} \tilde{x}) = \ln(E\{\tilde{x}^{T} \tilde{x}\}) + 1 = \ln\, (\mathrm{Tr}\{C_{x}\}) + 1 \tag{4.25}$$

Substituting (4.25) into (4.24) the entropy associated with the system $H(V_{lb})$ is decoupled in two terms

$$H(V_{lb}) = \ln\lambda_{max} + H(\tilde{x}^{T} \tilde{x}) \tag{4.26}$$

The first term, $\ln\lambda_{max}$, is a task dependent measure, characterizing the allowable uncertainty in the task description. The second term, $H(\tilde{x}^{T} \tilde{x})$, is a measure of the uncertainty contained in the state description of the system, and is independent of the proposed task. The importance of these results are that the entropy (uncertainty) can be reduced either by

- Decreasing ${}_{max.}$ implying relaxation of constraints l_i
- Decreasing the state uncertainty by improving sensing and control

Using again the Chebychev inequality the reliability of the system can be established

$$R_{lb} = 1 - E\{V_{lb}\} = 1 - \lambda_{max} E\{\tilde{x}^{T} \tilde{x}) = 1 - C_{lb} \exp H(\tilde{x}^{T} \tilde{x}) \tag{4.27}$$

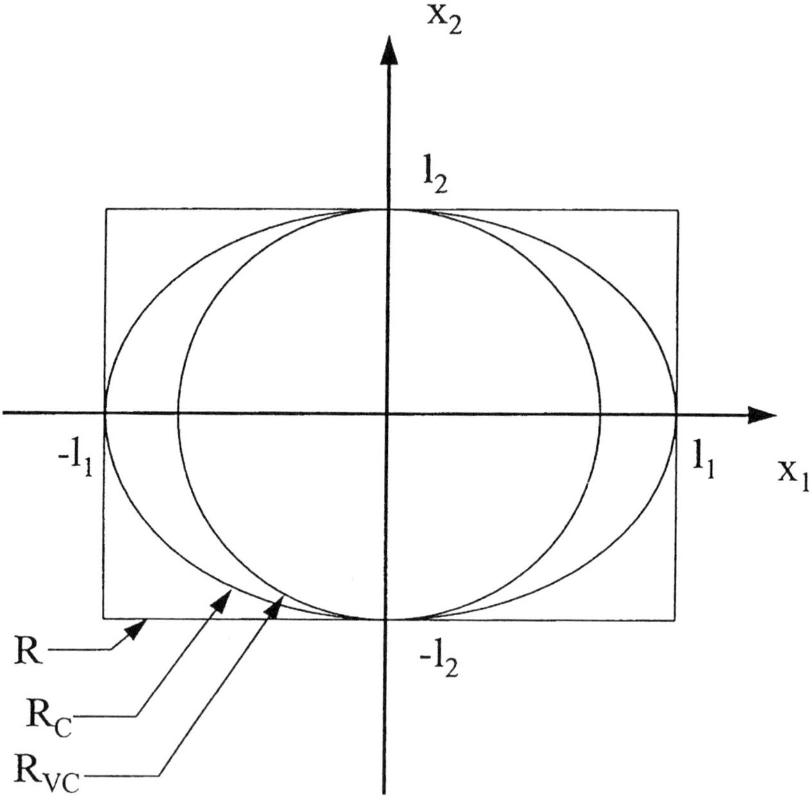

Fig. 4.2 Actual, Lower, and Loose Bound Reliability

$$C_{lb} = {}_{max.}\, e^{-1} \tag{4.28}$$

Fig. 4.3 COMPARISON OF ENTROPY MEASURES

Attribute	Lower Bound Method	Loose Bound Method
Cost Function	$V = \tilde{x}^T Q \tilde{x}$	$V_c = {}_{max.}\, \tilde{x}^T \tilde{x}$
Entropy function	$H(V)=\ln E\{V\}+1$	$H(V_{lb}) = \ln\lambda_{max}+H(\tilde{x}^T \tilde{x})$
Reliability	$R_c = 1\text{-}\exp H(V)/e$	$R_c = 1\text{-}\exp H(\tilde{x}^T \tilde{x})/e\lambda_{max}$
Accuracy	Tight Bound	Loose Bound
Coupling	States, Task coupled	States, Task decoupled
Use	Accurate Analysis	Simplified Design

Equations (4.27) and (4.28) imply that

- The loose bound reliability of the system, representing a strict bound, decreases exponentially with the increase of the entropy associated with the system states.
- The rate at which the reliability decreases is a function of the task description, $_{max.}$

The advantages of the entropy formulation are

- Analysis requires minimal statistical information
- No multidimensional p.d.f. 's must be developed or integrated
- Entropy-based systems are reinterpreted in terms of reliability
- Entropy function can be decoupled

Finally the following theorem applies to the analysis procedures

Theorem 4.3 The entropy function is invariant to homogeneous transformations

4.6 RELIABILITY-BASED INTELLIGENT CONTROL

The Hierarchically Intelligent Control problem was formulated in Chapter 3 using the minimization of an abstract cost function. The problem of optimal selection of tasks may be reformulated using the entropy-based reliability as the performance cost (McInroy, Musto, and Saridis 1995).

It is assumed that a command C is applied to the organization level and creates a sequence of subtasks s_i composing the respective task S:

$$S = \{s_1, s_2, \dots {}_{s.n.}\} \qquad (4.29)$$

This task description S is then communicated to the coordination level, in conjunction with a database of low level control, expressed in the system state variables x, to generate a set of feasible control alternatives to execute the task S. The set of m feasible plans is denoted by Q;

$$Q = \{Q_1, Q_2, \dots {}_{QM}\} \qquad (4.30)$$

where each plan Q_i is composed of subclass q_{ij}, associated with the n subtasks of S;

$$Q_i = \{q_{i1}, q_{i2}, \dots q_{in}\} \qquad (4.31)$$

Each subplan q_{ij} is composed of a k-tuple representing a unique combination of system design alternatives for the execution of the given subtask s_i. A cost $V_j = \tilde{x}^T R_j\, \tilde{x}$ is now assigned to each of the subtasks s_j, j=1,..n, which may be used to evaluate the entropy $H(V_{ij})$ describing the uncertainty of performance of each of the subclass q_{ij} in meeting the constraints of subtask s_j. A search is performed over the space of all Q_i to determine the minimum entropy/maximum reliability plan corresponding to the command C. Then the optimal plan P^* is scheduled and dispatched to the execution level for implementation.

The optimal plan, P^*, is implemented at the execution level of the system and its results are fed back to the coordination level to update its database. Specifically, the state variable error statistics are updated and direct learning of reliability information is performed. This learnt reliability information is then fed back to the organization level to update future command decomposition decisions. This **reliability assessment method** evaluates the low level plans at the coordination level by minimizing the system entropy as well as it maximizes a lower bound of the system reliability.

4.7 ILLUSTRATIVE EXAMPLE

The following example of a PUMA 560 robotic manipulator, performing high precision assembly, is used to demonstrate the reliability assembly method (Musto, Saridis 1993). The task $S=\{s_1\}$ is defined as positioning a part such that it can be inserted in a receptacle. Using the assembly tolerances in terms of a nominal set of position coordinates (x,y,z), and a set of bidirectional error constraints (δx,δy,δz) this task can be formulated. The single subtask of task S is defined by the tolerance constraints:

$$-0.0020 \le \delta x \le 0.0020 \text{ m}$$
$$-0.0020 \le \delta y \le 0.0020 \text{ m}$$
$$-0.0025 \le \delta z \le 0.0025 \text{ m}$$

The task should be successfully completed if the above specifications are met. Assume the following constraint matrix as in section 4.5:

$$R_1 = \begin{bmatrix} 250{,}000 & 0 & 0 \\ 0 & 250{,}000 & 0 \\ 0 & 0 & 160{,}000 \end{bmatrix}$$

In order to complete this task, the Intelligent Control system must select, from available alternatives, a plan involving an inverse kinematic solution and a set of control gains that maximizes the reliability of the system with respect to the proposed task

Based on the nominal position specification associated with s_1, three manipulator configurations are capable of reaching the desired point, e.g.;

$$\theta_1 = [-90^0 \quad -10.14^0 \quad 159.23^0 \quad 0^0 \quad -14.08^0 \quad 90^0]^T$$
$$\theta_2 = [115.27^0 \quad -169.86^0 \quad 26.16^0 \quad 72.22^0 \quad 19.06^0 \quad -142.86^0]^T$$
$$\theta_3 = [115.27^0 \quad -169.86^0 \quad 26.16^0 \quad -107.78^0 \quad 19.06^0 \quad 37.14^0]^T$$

In order to reach this point in the (xyz) space, two possible feedback control algorithms are available both of the computer torque category:

$$\tau = D\,[\ddot{\theta}_d + K_v(\dot{\theta}_d - \dot{\theta}) + K_p(\theta_d - \theta)] + C$$

where D is diagonal constant estimate of the mass matrix, $\dot{\theta}$ and θ are noisy joint velocity and position measurements; $\ddot{\theta}_d$, $\dot{\theta}_d$, and θ_d are the reference joint acceleration, velocity, and position; C is an approximation of the gravity term. They differ in terms of their gains;

C_1: $K_p = -100$, $K_v = -20$
C_2: $K_p = -75$, $K_v = -15$

Therefore, there are six alternative plans to be evaluated by the Intelligent Control system, each described by:

$$Q_k = \{ C_i, \theta_j \} \quad k = 1, \ldots 6$$

Note, there is only subplan q_{k1} in each plan Q_k .

In order to select the optimal plan Q^*, statistical estimates of the performance of the plans were obtained through simulations with added noise. These are given in Table 4.1. Using the lower bound formulation an entropy value and the corresponding lower bound and reliability estimates are calculated for each plan and are given in Table 4.2 along with simulated actual reliability values. Plan Q_2 is shown to be the optimal.

The alternative loose bound formulation was also used for comparison purposes and

Table 4.1 Position Error Variances

Plan Number	Plan Description	$\sigma^2_{\delta x}$	$\sigma^2_{\delta y}$	$\sigma^2_{\delta z}$
P_1	$\{C_1,\Theta_1\}$	0.0925	0.0439	0.1552
P_2	$\{C_1,\Theta_2\}$	0.0912	0.0243	0.1330
P_3	$\{C_1,\Theta_3\}$	0.0932	0.0341	0.1355
P_4	$\{C_2,\Theta_1\}$	0.0854	0.0555	0.1414
P_5	$\{C_2,\Theta_2\}$	0.0732	0.0657	0.1122
P_6	$\{C_2,\Theta_3\}$	0.0848	0.0657	0.1510

(all values $\times 10^{-5}$)

Table 4.2 Lower Bound Analysis

Plan Number	$H(V)$	R_{lb}	Ranking	Reliability
1	0.4712	0.4107	5	0.8167
2	0.3099	0.4985	1	0.8638
3	0.3746	0.4650	3	0.8457
4	0.4527	0.4215	4	0.8225
5	0.3590	0.4732	2	0.8502
6	0.5185	0.3822	6	0.8018

Table 4.3 Loose Bound Analysis

Plan Number	$H(V_c)$	R_{vlb}	Ranking	Reliability
1	0.6839	0.2710	5	0.8167
2	0.5240	0.3788	1	0.8638
3	0.5799	0.3430	3	0.8457
4	0.6515	0.2942	4	0.8225
5	0.5344	0.3723	2	0.8502
6	0.7173	0.2462	6	0.8018

shown in Table 4.3. Plan Q_2 is again selected as the optimal plan. The rankings of the calculated plans are shown to be the same in both formulations

This example the simplicity of the use of entropy measures for optimizing Intelligent Control problems. It also demonstrates the usefulness of reliability to improve the performance of Hierarchically Intelligent Control.

4.8 CONCLUSIONS

The interpretation of reliability as an entropy and the further analytic formulation of the subject was introduced as an example of the use of entropy in scientific and engineering applications. It bears only indirect relation to the control problem by evaluating the cost of

performance of Intelligent Control systems, their task selection, the acceptance or not of a specific control activity and other similar situations. In the process of doing so valuable analytic simplifications like the looser lower bound were developed that gave better insight of the expression of reliability. It also simplified its calculation at the cost of tighter acceptance conditions. Reliability, of course applies to all fields of engineering and this approach may prove valuable to all of them

4.9 REFERENCES

Harr M.E. (1987), ***Reliability-Based Design in Civil Engineering*** McGraw-Hill New York NY.

McInroy J.E., Saridis G.N.,(1991), "Reliability Based Control and Sensing Design for Intelligent Machines", in *Reliability Analysis* ed. J.H. Graham, Elsevier North Holland, N.Y.

McInroy J.E., Musto J., Saridis G.N. (1995) ***Reliable Plan Selection by Intelligent Machines*** World Scientific, Singapore.

Musto J., Saridis G.N. (1993), "An Entropy-Based Assessment Technique for Intelligent Machines" *Proceedings 1993 Conference on Intelligent Controls,* Chicago IL, Aug.

Saridis G.N. (1995), "Architectures of Intelligent Controls" *Intelligent Control Systems,* M. M. Gupta, N. Sinha eds., IEEE Press.

CHAPTER 5

ENTROPY IN INTELLIGENT MANUFACTURING

5.1. AUTOMATION

The evolution of the digital computer in the last thirty years has made possible to develop fully automated systems that successfully perform human dominated functions in industrial, space, energy, biotechnology, office, and home environments, generating waste interpreted as entropy. Therefore, automation has been a major factor in modern technological developments. It is aimed at replacing human labor in

a. hazardous environments,
b. tedious jobs,
c. inaccessible remote locations and
d. unfriendly environments.

It possesses the following merits in our technological society: reliability, reproducibility, precision, independence of human fatigue and labor laws, and reduced cost of high production.

Modern Intelligent Robotic Systems, using entropy as a measure of performance, are typical applications of Automation to an industrial society (Valavanis, Saridis 1992). They are equipped with means to sense the environment and execute tasks with minimal human supervision, leaving humans to perform higher level jobs.

Manufacturing on the other hand, is an integral part of the industrial process, and is defined as follows:

> **Manufacturing is to make or process a finished product through a large scale industrial operation.**

In order to improve profitability, modern manufacturing, which is still a **disciplined art**, always involves some kind of automation. Going all the way and fully automating manufacturing is the dream of every industrial engineer. However, it has found several roadblocks in its realization, measured by entropy: environmental pollution, acceptance by the management, loss of manual jobs, marketing vs. engineering. The National Research Council reacted to these problems by proposing a solution which involved among other items a new discipline called: **Intelligent Manufacturing** (The Comprehensive Edge 1989).

Intelligent Manufacturing is the process that utilizes Intelligent Control, with entropy as a measure, in order to accomplish its goal. It possesses several degrees of autonomy, by demonstrating (machine) intelligence to make crucial decisions during the process. Such decisions involve scheduling, prioritization, machine selection, product flow optimization, etc., in order to expedite production and improve profitability.

5.2. INTELLIGENT MANUFACTURING

Intelligent Manufacturing is an immediate application of Intelligent Control discussed in Chapter 3. It has been defined as the combination of disciplines of Artificial Intelligence, Operations Research and Control System Theory (see Fig. 5.1), in order to perform tasks with minimal interaction with a human operator. One of its hierarchical applications, proposed by Saridis (1996), is an architecture based on the **Principle of Increasing Precision with Decreasing Intelligence (IDI),** which is the manifestation on a machine of the human organizational pyramid The Principle is realized by three structural levels using entropy as a common measure (see Fig. 5.2):

1. The Organization level (Saridis, Moed 1988)
2. The Coordination level (Saridis, Graham 1984, Wang, Saridis 1990)
3. The Execution level (Saridis 1979).

Intelligent Manufacturing can be implemented in the Factory of the Future by modularizing the various workstations and assigning Hierarchically Intelligent Control to each one of them, the following tasks;

1. Product Planning to the Organization level
2. Product Design and Hardware Assignment and Scheduling to the Coordination level
3. Product Generation to the Execution level

The algorithms at the different levels may be modified according to the taste of the designer, and the type of the process. However, manufacturing can be thus streamlined and optimized by minimizing the total entropy of the process. Robotics may be thought as an integral part of Intelligent Manufacturing and be included as part of the workstations. This creates an intelligently versatile automated industrial environment where, every time, each unit may be assigned different tasks by just changing the specific algorithms at each level of the hierarchy (see Fig. 5.3) in contrast to the serial production which requires more equipment and effort (Fig. 5.4).

This approach is designed to reduce interruptions due to equipment failures, bottlenecks, rearrangement of orders, material delays and other typical problems that deal with production , assembly and product inspection. A case study dealing with a nuclear plant may be found in (Valavanis, Saridis 1992).

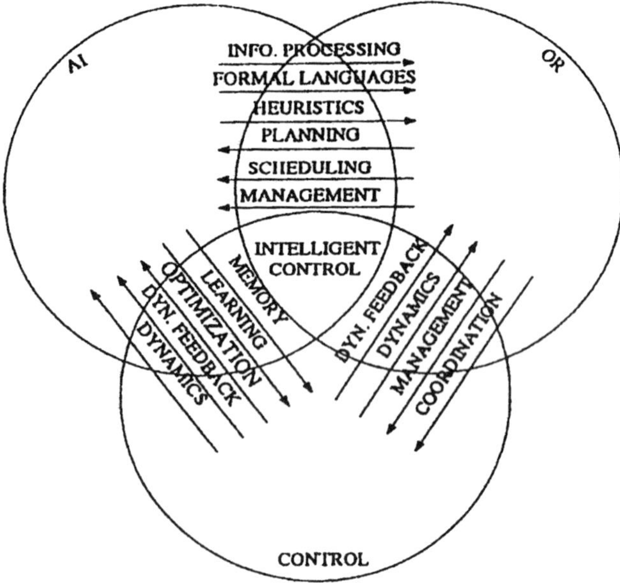

Fig. 5.1 Definition of Intelligent Control

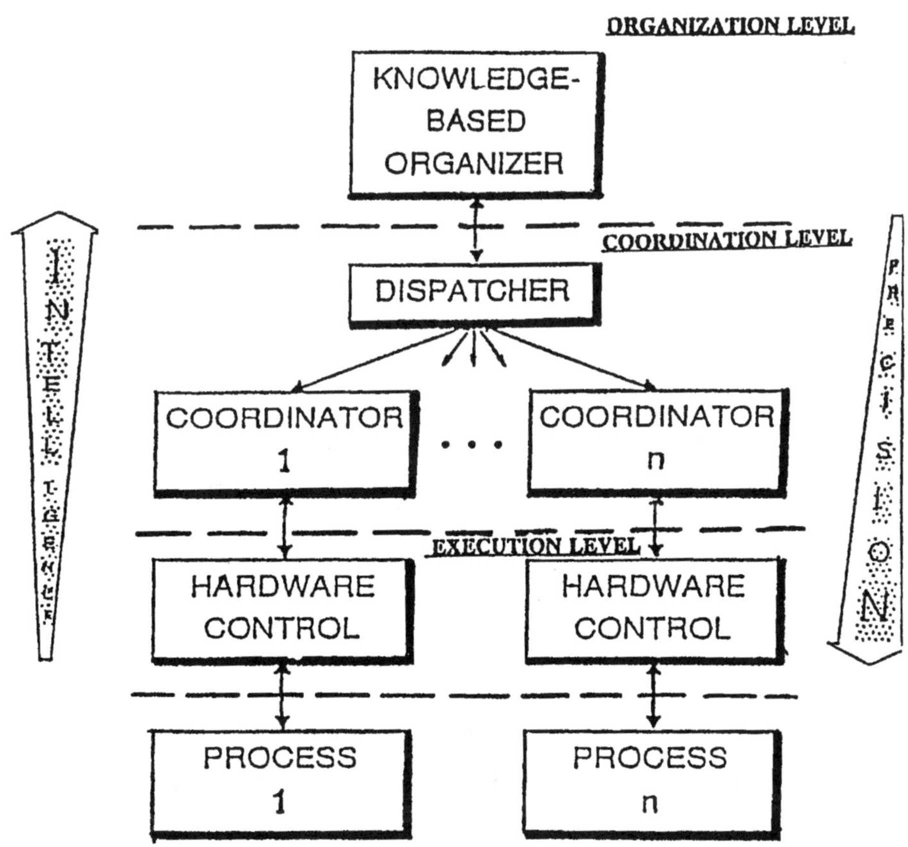

Fig. 5.2 Structure of Intelligent Machines

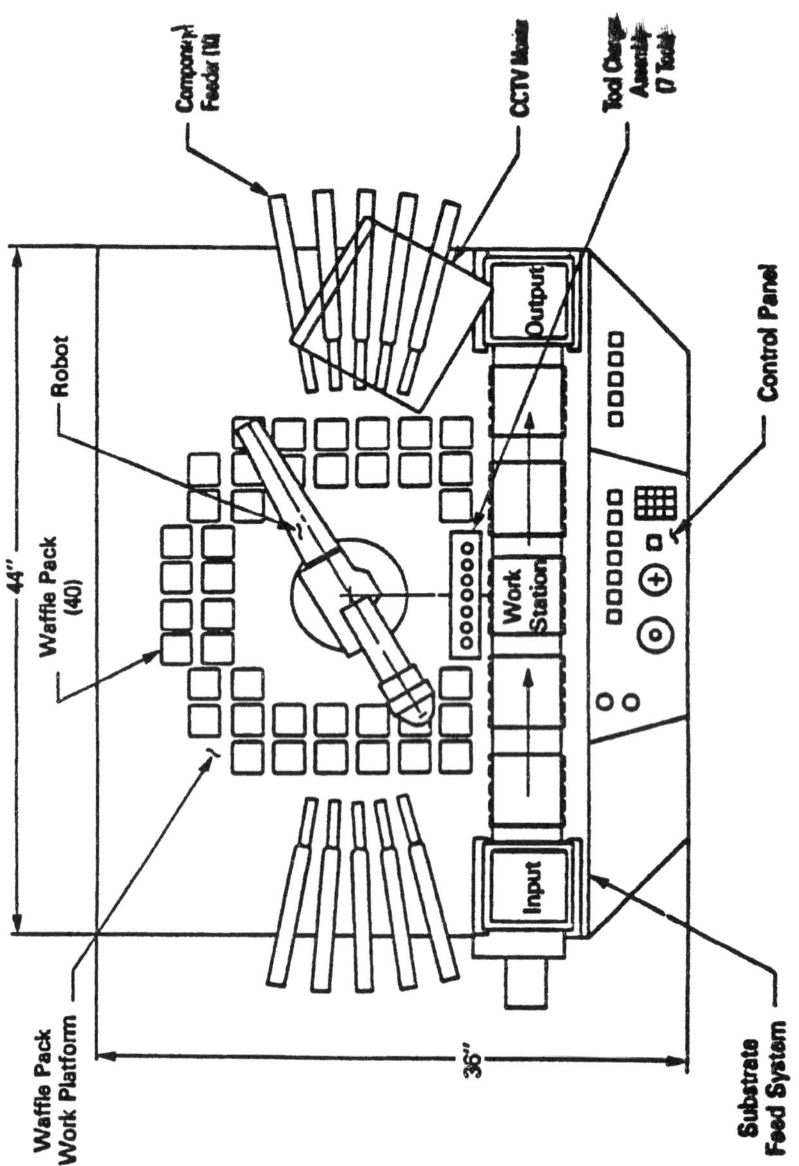

Fig.5.3 Circular Assembly System

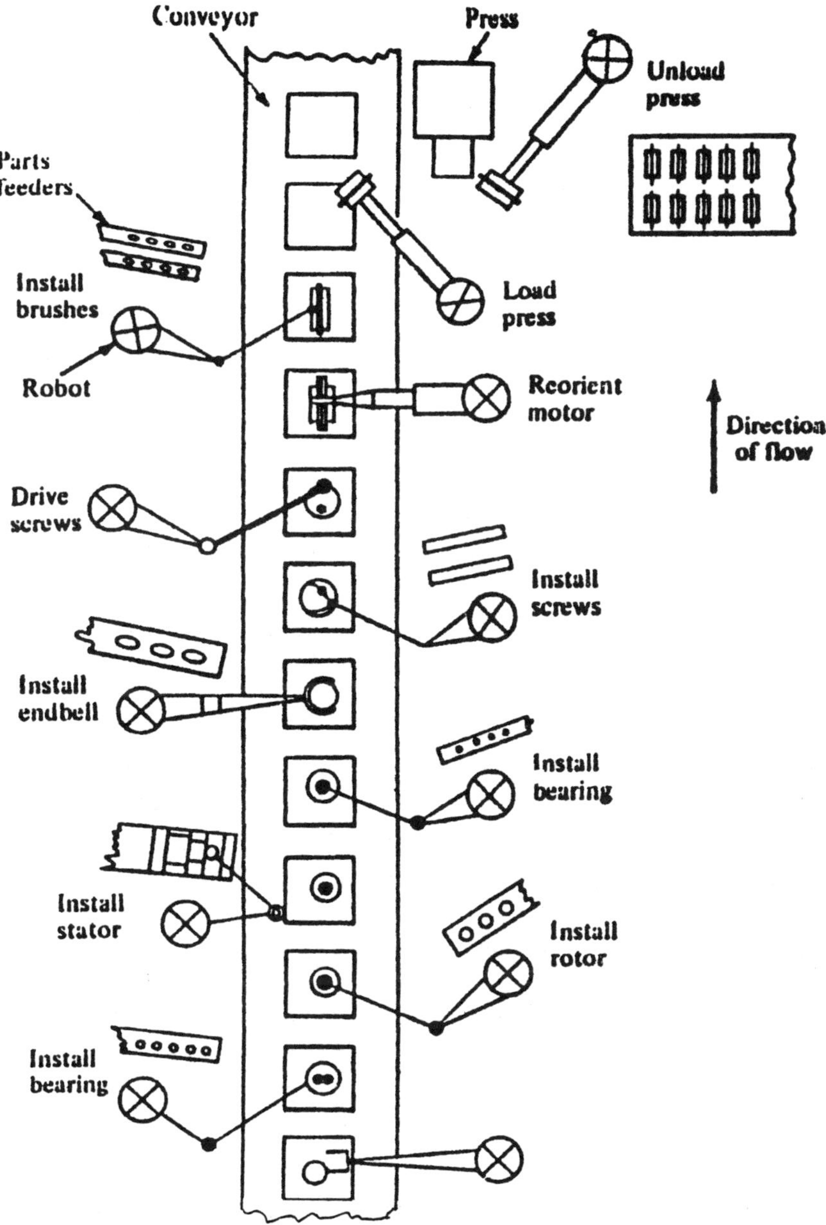

Fig 5.4 Series Assembly System

At the present time the application of such technology, even though cost-effective in competitive manufacturing, is faced with significant barriers due to (The Comprehensive Edge 1989);

a. Inflexible organizations
b. Inadequate available technology
c. Lack of appreciation and
d. Inappropriate performance measures

However, international competition, and need for more reliable, precisely reproducible products is directing modern manufacturing towards more sophistication and the concept of an Intelligent Factory of the future. An interesting application based on the work of Varvatsoulakis, Saridis, and Paraskevopoulos (1998) is discussed in the sequel.

5.3 INTELLIGENT PRODUCT SCHEDULING FOR MANUFACTURING

The theory of Hierarchical Intelligent Control is general and versatile enough to be used to control as well as schedule tasks in Manufacturing. Such is the case of automated multiple product scheduling in a modern factory.

Automated multiple product scheduling is needed when the factory produces more than one product on the same set of stations and the ordering of production must be set as a minimum operating cost scheduling problem. The problem is mathematically formulated to set the order of production using entropy as a measure in the Intelligent Control's three level structure (Varvatsoulakis, Saridis, and Paraskevopoulos 1998, 1999). The complete system is able to issue high-level task plans and use them to control the stations of the factory in the lower level of the hierarchy. The system includes a learning algorithm designed to obtain asymptotically optimal task plans for production control in uncertain environments. A case study in (Varvatsoulakis, Saridis, and Paraskevopoulos 1999) demonstrates the scheduling of the assembly of a gear box. The intelligent manufacturing approach is presented in the following sections.

5.3.1 Product Scheduling Architecture: The Organization Level

The Organization level, discussed in Chapter 3, is intended to perform operations as *Task Representation, Planning, Decision Making, and Learning*, in an abstract way and produce strings of commands to be executed by the lower levels of the system. A Boltzmann type of neural net was found to suffice to produce these operations. The analytic model proposed for multiple product scheduling in manufacturing follows the pattern of hierarchically intelligent control and is divided in three levels.

The organization level, (Fig. 5.5), that sets up the task planning is composed of the following subtasks:

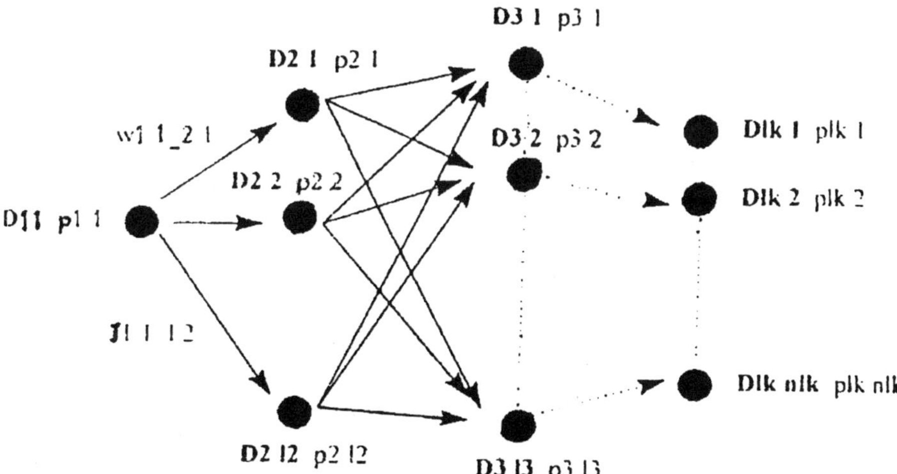

Fig. 5.5 Boltzmann Machine Representation

Task Representation is the association of a station node to a number of tasks. The probability of its activation, and the weight for transfer are assigned and its associated entropy is calculated. Define:

- The ordered set of sublevels $L=\{l_1,...l_k\}$ is the set of abstract primitive subtasks, each

one containing number nl_i, $l=1,...k$ of independent primitive nodes.

- The set of nodes $D=\{d_{i1},...d_{i,nl};\ l=1,...k\}$, is the subtask domain with each node containing number of primitive objects; for convenience they are represented by their subscripts.
- The set $B\subseteq D$ contains the starting nodes.
- The set $S\subseteq D$ contains the terminal nodes
- The set of random variables $Q=\{q_{i1},...q_{i,nl};\ l=1,...k\}$ with binary values [0,1], represents the inactive or active state of events associated with the nodes D.
- The set of probabilities $P=\{p_{i1},...p_{i,nl};\ l=1,...k\}$ associated with the random variables Q, as follows:

  $P=\{p_{ij}=Prob(q_{ij}=1);\ l=1,...k;\ j=1,...nl$

Task planning is the ordering of the production activities, obtained by properly concatenating the appropriate abstract primitive nodes $\{d_{m,li};\ m=1,..nl_i\}$, for each sublevel l_i. The ordering is generated by a Boltzmann machine, which represents the whole organization level, and measures the flow of knowledge $R_{im,i+1i}$ from node of sublevel l_i to node l_{i+1},

$$R_{im,i+1j} = -\tfrac{1}{2}\, w_{im,i+1i}\, q_{im}\, q_{ii+1j} \tag{5.1}$$

where the connection weight $w_{im,i+1j} > 0$

The transition probability is calculated to satisfy Jaynes' maximum principle:

$$p_{im,ii+1j} = \exp[-a_{im} - \tfrac{1}{2}w_{im,i+1j}\, q_{im}\, q_{ii+1j} \qquad \sum_{J=1}^{nll+1} p_{im,ii+1j} = 1 \tag{5.2}$$

with a_{im} an appropriate normalization constant.

The negative entropy, in Shannon's sense, transferred from node d_{im} to d_{i+1j}, is

$$H_{im,i+1j} = -E\{\ln p_{im,ii+1j}\} = \hat{a}_{im} + \tfrac{1}{2}w_{im,i+1j}\, q_{im}\, q_{ii+1j} \tag{5.3}$$

where $\hat{a}_{im}$ is an appropriate normalization constant. $H_{im,i+1i}$, is an increasing function of probabilities and weights, and defines the order of concatenation of the nodes.

<u>Decision Making</u>, is associated with the Boltzmann Machine, and is obtained as follows:

Starting at the node d_{im} the connections to nodes d_i+1j are searched until node j^*, corresponding to the node of minimum entropy is found.

$$j^* = \arg\max_j H_{im,i+1j} = \arg\max_j (\hat{a}_{im} + \tfrac{1}{2} w_{im,i+1j}\, q_{im}\, q_{ii+1j})$$
$$H_{im,i+1j}{}^* = [H_{im,i+1j}]_{j=j^*} \tag{5.4}$$

<u>Learning (Feedback)</u> is obtained by upgrading the probabilities and weights of the successful path after each iteration. A double stochastic approximation recursive algorithm is used for the upgrading. The upgrade of the estimate of the cost function between nodes d_{im} and d_{i+1j} ,namely $J_{im,i+1j}(t)$, is given at the (t+1)st iteration, by:

$$J_{im,i+1j}(t+1)= J_{im,i+1j}(t) + (t+1)^{-1}[J_{obs}(t_k+1) - J_{im,i+1j}(t)] \tag{5.5}$$

where $J_{im,i+1j}(t)$ is the performance estimate, J_{obs} is the observed value and

$$p_{im}(t+1) = p_{im}(t) + (t+1)^{-1}[\bar{p} - p_{im}(t)] \tag{5.6}$$
$$w_{im,i+1j}(t+1) = w_{im,i+1j}(t) + (t+1)^{-1}[\bar{w} - w_{im,i+1j}(t)]$$

$$\bar{p}, \bar{w} = \Gamma_k(t_k+1) = \begin{cases} 1 & \text{if } J = \min \Sigma J \\ 0 & \text{otherwise} \end{cases} \tag{5.7}$$

All these operations are performed by the Boltzmann Machine presented in Chapter 3, and shown in Fig. 5.5.

5.3.2 Product Scheduling Architecture: The Coordination Level

The Coordination level (Varvatsoulakis et. Al. 1999) serves in this case as the interface between the Organization and the Execution levels. It maps the abstract tasks into the real world, without involving data packages left for the Execution level. Its *dispatcher* interprets and decomposes the commands received by the Organization level and distributes them to the appropriate coordinators which correspond to the stations of the factory. They in turn translate the control commands into operation instructions for the execution devices.

The class of coordinators, considered here, is an extension of deterministic finite state machine generators, representing a discrete event dynamic system that evolves according to the occurrence of spontaneous event generating state transitions (Ramage, Wonham 1987).

The associated finite state space is generated by the n-dimensional state variables X defined over a set SX.

$$X = [x_1,...x_n]\ ;\ \ x_i \in SX, \quad I=1,...n \qquad n\in N \tag{5.8}$$

Definition 5.1 A Finite State Machine Generator (FSMG) is a sextuple (X,U,f,g,X_0,X_f) where (Fig. 5.6)

X is the finite state space
U is the alphabet of events
Y is the output alphabet
$f{:}XxU \rightarrow X$ is the transition function
$g{:}XxU \rightarrow Y$ is the output function
$X_{ij} \in X$ is the initial state
$X_f \subseteq X$ is the set of final states representing completed tasks.

The dynamics of the FSMG is described by the difference equation:

$$\begin{aligned} X(k+1) &= f(X(k), U(k)) \\ Y(k+1) &= g(X(k+1), U(k)); \end{aligned} \tag{5.9}$$

where $X(k+1) \in X$, the state after the kth event; $U(k) \in U$, the kth event; and $Y(k+1) \in Y \subset 2^Y$ the set of all possible output symbols.

In terms of a formal regular language let:

$U^* \subseteq U$ denote the set of all finite strings including the empty string e, and the sequence at instance k, $U^*(k)=\{U(k),...U(0)\}$. Letting $f{:}XxU \rightarrow X$ be extended to a function, the internal behavior of FSMG is described by a formal language, denoting the traces of events accepted by FSMG:

$$L_x(FSMG) \subseteq U^* := \{u^* \in U^* \mid f(u^*, X_0) \text{ is defined}\} \tag{5.10}$$

The Language L_{xf} of all finite traces representing the complete tasks generated by FSMG is defined as:

$$L_{xf}(FSMG) \subseteq L_x(FSMG) := \{\, u^* \in L_x(FSMG) \mid f(\, u^*, X_0) \in X_f\} \tag{5.11}$$

In a similar way let Y^* of all finite strings over Y including the empty string e and $Y^*(k)=\{Y(k),...Y(0)\}$containing the output symbols at time k. Letting $h{:}Y^*xX \rightarrow X$, the output behavior of FSMG is described in terms of the formal regular language, denoting all finite

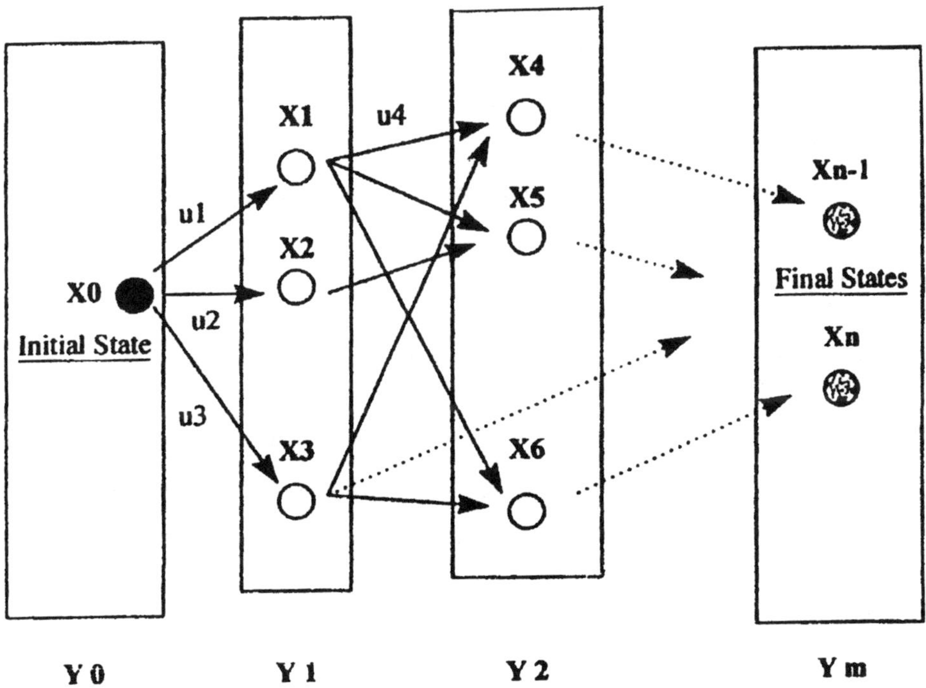

Fig. 5.6 Finite State Machine Representation

traces of output symbols generated by FSMG.

$$L_Y(FSMG) \subseteq Y^* := \{y^* \in Y^* \mid y^* = h(u^*, X_0) \text{ is defined}, u^* \in U^*\} \quad (5.12)$$

The Language L_{Yf} representing the finite traces of output symbols complete tasks generated by FSMG is defined as:

$$L_{Yf}(FSMG) \subseteq L_Y(FSMG) := \{ y^* \in L_Y(FSMG) \mid h(y^*, X_0) \text{ is defined}, f(u^*, X_0) \in X_f\} \quad (5.13)$$

A sequence of states {X(k),...X(0)} is called a *path* of FSMG if there exists a sequence of event symbols {U(k),...U(0)} such that X(k+1)=f(X(k)U(k)), k=1,...n-1. When X(k)= X(0) this path is called a*cycle*. If an FSMG has no cycles it is called *acyclic*, which is the case of our coordinators.

In the linguistic decision approach, task processes of a FSMG are the processes of translation

$$L_x(FSMG) \to L_y(FSMG) \quad \text{and}$$
$$L_{xf}(FSMG) \to L_{yf}(FSMG) \quad (5.14)$$

The formal languages L_x(FSMG) and L_{xf}(FSMG) define the control languages for a level, which through FSMG generate the set of control commands for the lower level.

In standard control terminology FSMG plays the role of the open loop plant with a certain physical behavior. The control mechanism available to the supervisor is the ability to block the occurrence of some events, in order that the whole system meets certain specifications

Definition 5.2. An FSMCG = (X,U,f,g,h,X_0,X_f) is a *finite state machine control generator* where

X is the finite state space
U is the alphabet of events
Y is the output alphabet
$f: XxU \to X$ is the transition function
$h: XxU \to Y$ is the output function
$g:: XxU \to \{0,1\}$ is the control function
$X_0 \in X$ is the initial state
$X_f \subset X$ is the set of final states representing completed tasks.

The behavior of the FSMCG is described by:

$$X(k+1) = f(X(k), U(k)) \qquad \text{if } g(X(k), U(k)) = 1$$
$$X(k+1) = X(k) \qquad \text{if } g(X(k), U(k)) = 1$$

$$Y(k+1) = h(X(k+1), U(k)); \quad \text{if } g(X(k), U(k)) = 1$$
$$Y(k+1) = Y(k) \quad \text{if } g(X(k), U(k)) = 0 \qquad (5.15)$$

Control of an FSMCG consists of switching the control function g(X(k),U(k)) according to defined conditions. If g(X(k),U(k))=0 the system is blocked and no state transition is allowed . This way every new event symbol U(k) is rejected and not recorded in the event string. This controller is *static*.

Letting g: $X^* \times U^* \to \{0,1\}$, the extended control function $g(X^*,U^*)$ depends on full state and event string observation. This controller is *dynamic*.

5.3.3 Product Scheduling Architecture: The Execution Level

The Execution level constitutes of the workstations corresponding one-to-one to the appropriate coordinators. Their effort is measured with a cost function interpreted as entropy, and fed back to the coordinators to make the selection of the proper path in product scheduling . They are part of the factory's hardware.

5.4.. AUTOMATED PRODUCT SCHEDULING: A PARADIGM.

A model of the production scheduling process is presented, for a particular class of problems, by the assembly of a gear box, from four different parts represented by the small square boxes in Fig. 5.7. It is based on the previously presented hierarchically intelligent product scheduling procedure.

5.4.1 The Organization Level Structure

The main function of the Organization level is to construct a set of complete strings of events, that represent various tasks, as it was previously stated. It deals with the four operations ;

- Task representation
- Task Planning
- Decision Making
- Learning (Feedback)

A Boltzmann machine, described in section 5.3.1, can provide completely the optimal task representation, planning, decision making and learning of the sequence of events based on uncertain measurements. According to the specified it is possible to dynamically eliminate nodes or connections by properly defining $\bar{p}$, $\bar{w}$.

More particularly, for the structure of Fig. 5.7, the following sets are defined:

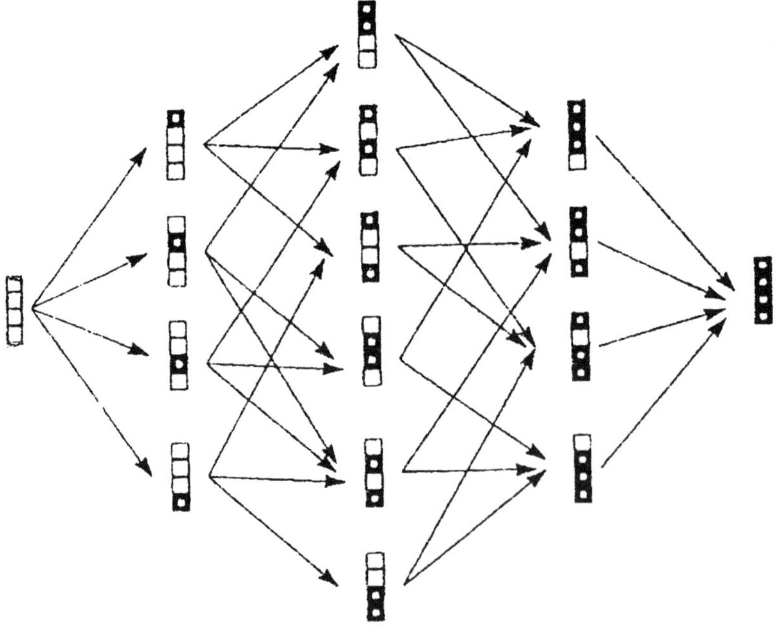

Fig. 5.7 Alternative Assembly Schedules

- $L = \{l_i, l=1,...n\}$ the ordered set of levels, with n the number of assembly parts.
- $D = \{d_{i1},...d_{in}; l=0,...n\}$ the set of nodes representing the parts assembled at each level l_i, where the number of nodes is n!/(n-l)!l!.
- U = {fetch part l; l=1,...n} the set of events, transferring between nodes.
- The set P of node probabilities, and the set W of the transition weights between nodes. A Memory storage area to back up the probabilities associated with organizer during the intermediate and final phase of the learning process is a necessary device to be part of the Organization level.

A search maximize the entropy between nodes is performed to generate the optimum string of the assembly procedure. After each iteration the cost of task execution is fed back from the Coordination level to upgrade the probabilities and the transition weights.

5.4.2 The Coordination Level Structure

The Coordination level is composed of a fixed *dispatcher* and several flexible *coordinators*. The dispatcher , modeled by an FSMG, receives from the Organizer the sequences of

events as a string of commands and translates them into control actions for the corresponding Coordinators as control actions(Fig. 5.7). Each Coordinator, modeled as a Finite State Machine Control Generator (FSMCG), uses the control actions to produce instructions transmitted to a *communication bus* for the initialization of other coordinators Let $SX = \{q_i; l=1,...n\}$ contain the parts of the assembly. If $m \leq n$ is the number of parts already assembled the state of FSMG is the vector $X_m = \{x_i; l=1,...m; x_i \in SX\}$, representing the parts assembled and $X_n = \{x_i=0; l=m+1,...n; x_i \in SX\}$. The set of events $U=q \in SX$ and $X(k)=\{X_m(k),X_n(k)\}$, define the function

$$X(k+1) = f(X(k), U(k))$$

while $U(k) = \{x_1,...x_m,x_{m+1},0,...0\}$ for $x_{m+1}=q$. A scheduling procedure for the coordinators can be designed based on the output function of FSMG. Symbols from the dispatcher are read by each coordinator through the communication bus. They switch the control function g(X(0),U(0)) from 0 to 1 to start the process; then it continues as long as g(X(k),U(k))=1 and blocks it when g(X(k),U(k))=0, expecting appropriate symbols from the other coordinators to switch back.

In this coordination structure, equivalent to the one proposed by Wang and Saridis (1990), the dispatcher has a dominant position, serving as the task control center and the only information communication channel among the coordinators.

After each execution the performance update is evaluated with the formula:

$$J_i(k+1)= J_i(k) + (k+1)^{-1}[J_{obs}(k) - J_i(k)] \qquad (5.16)$$

where $J_{obs}(k)$ is the measured performance at the kth iteration. The node probabilities are updated using the expression:

$$p_i(k+1) = p_i(k) + (k+1)^{-1}[p - p_i(k)]$$

$$p_i = \begin{cases} 1 & \text{if } J = \min J_i(k),\ l=1,\dots m \\ 0 & \text{otherwise} \end{cases} \qquad (5.17)$$

5.4.3 The Execution Level Structure

The Execution level consists of groups of devices corresponding one-to-one with the appropriate coordinators. The subcosts for each device are measured as entropies, after each complete execution of the task. Then they are transferred BACK to the coordination level for computation of the overall measured cost and upgrading of the upper levels.

5.5 SIMULATION RESULTS FOR THE ASSEMBLY OF A MACHINE

This special case study involves the simulation of a multilevel decision making system based on the assembly of four different parts of a machine described in Fig. 5.7. The black boxes in that figure represent assembled parts.

The Boltzmann machine for the Organization level is represented by four sublevels L==(l_1, l_2, l_3, l_4) and nl_1=1, nl_2=4, nl_3=6, nl_4=4, giving 24 total number of schedules. The maximum entropy schedule is transmitted to the Coordination level (see Fig.5.8).

There are four nodes D and the total number of paths, representing alternate production plans, is 24. All weight and probability limits are equal to 1. The costs of all paths are observed and estimated during every iteration. The optimal schedule is represented with the bold lines in Fig. 5.8.

The minimum estimated cost in terms of number of iterations is given in Fig.5.9. The maximum number of iteration to converge to the optimal sequence of a defined path is about 250.

5.6 CONCLUSIONS

After the presentation of the modern industrial setups, the impact of entropy on modern Intelligent Manufacturing. In particular the problem of scheduling of production in a modern factory was tackled, using a hierarchically intelligent control formulation using entropy as a measure of the cost of production, developed in Chapter 3 for Intelligent Machines. Other applications of the same concept can be and should be developed for the "Factory of the Future".

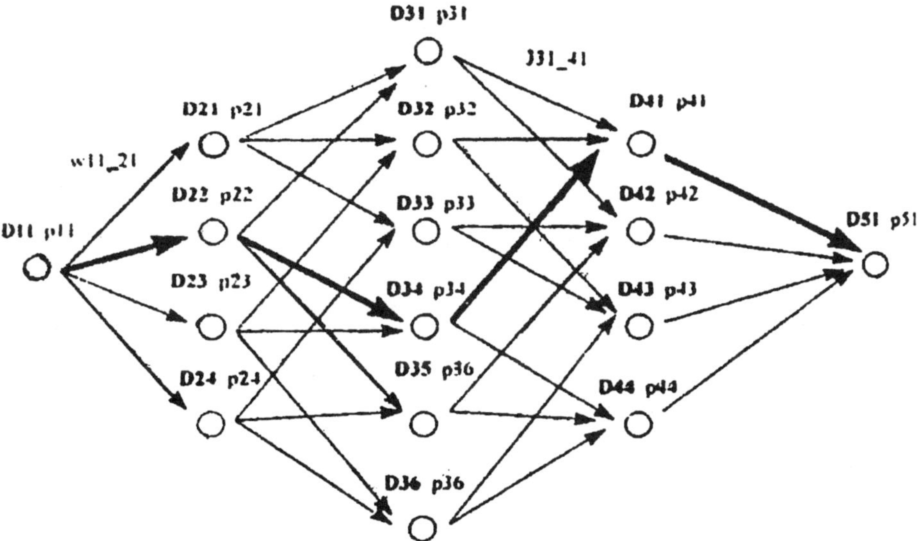

Fig. 5.8 Organization Level of a Manufacturing Network

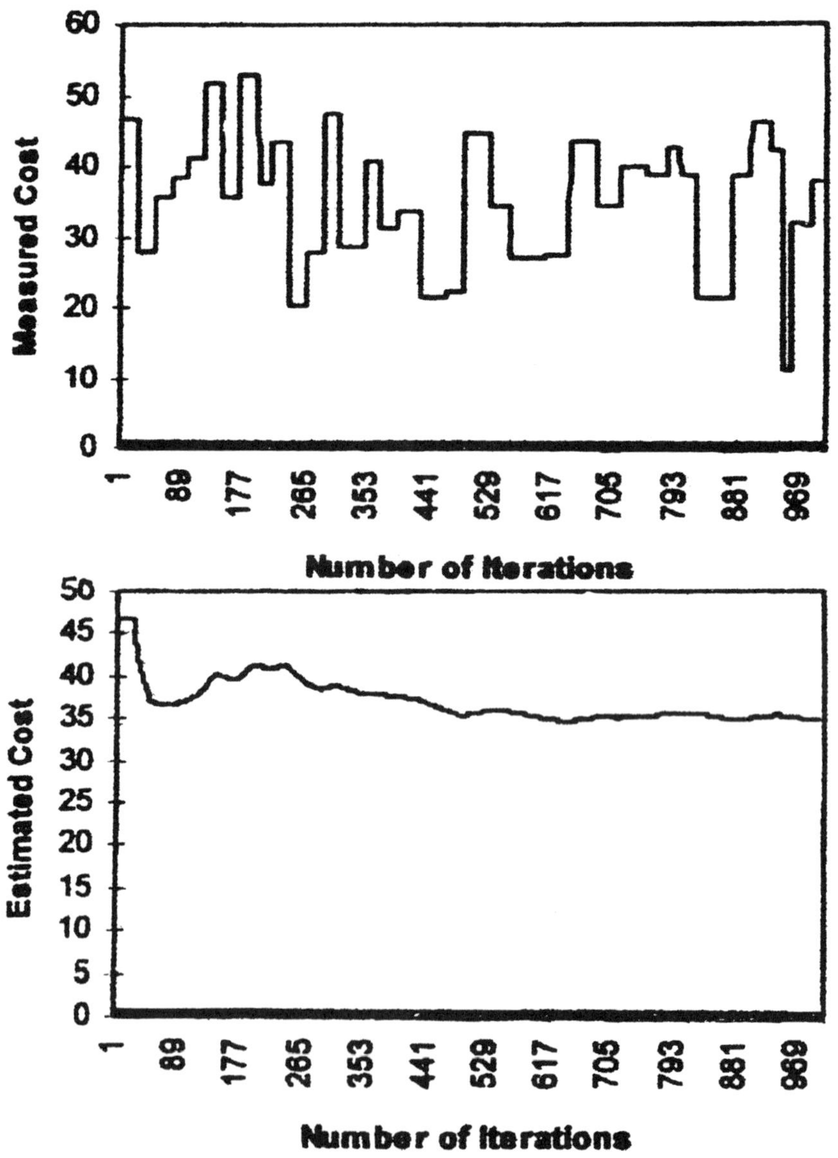

Fig.5.9 Estimated Assembly Cost (Entropy)

In this Chapter the flexibility of the Intelligent Control architecture was demonstrated. The Petri Net formulation developed in Chapter 3 for the Coordination level was replaced by a Discrete-event system, using always entropy as the cost of performance of the system. Other suitable methodologies may also be applicable, and the most efficient one should be selected for the appropriate problem.

5.7 REFERENCES

The Competitive Edge: Research Priorities for U.S. Manufacturing, (1989) ***Report of the National Research Council on U.S. Manufacturing,*** National Academy Press.

Ramage P. J., and Wonham W. M., (1987), "Supervisory Control of a class of discrete event processes" ***SIAM Journal of Control Optimization,*** Vol. 25, No. 1, pp.206-230

Saridis G. N. (1996)"Architectures for Intelligent Controls" in ***Intelligent Control Systems***, Edited by Madan M. Gupta and Naresh K. Sinha, Chapter 6, pp.127-148, IEEE Press.

Saridis G. N. (1979), "Toward the Realization of Intelligent Controls", *IEEE Proceedings,* ***67,*** No. 8.

Saridis, G.N. and Graham, J.H. (1984), "Linguistic Decision Schemata for Intelligent Robots", *AUTOMATICA the IFAC Journal,* ***20,*** No. 1, pp. 121-126, Jan.

Saridis, G.N. and Moed, M.C. (1988), "Analytic Formulation of Intelligent Machines as Neural Nets", ***Proceedings Symposium on Intelligent Control,*** Washington, D.C., August.

Wang, F., Saridis, G.N.(1990) "A Coordination Theory for Intelligent Machines" ***AUTOMATICA the IFAC Journal,*** *35*, No. 5, pp. 833-844, Sept.

Valavanis, K.P., Saridis, G.N.(1992), ***Intelligent Robotic System Theory: Design and Applications,*** Kluwer Academic Publishers, Boston MA.

Varvatsoulakis M., Saridis G. N., and Paraskevopoulos P., (1998) "A Model for the Organization Level of Intelligent Machines" ***Proceedings of 1998 International Conference on Robotics and Automation,*** Leuven Belgium May 15.

Varvatsoulakis M., Saridis G. N., and Paraskevopoulos P., (1999) "Intelligent Structures for Automated Assembly Scheduling" Submitted to ***IEEE Transactions on Control Systems Technology.***

CHAPTER 6

ENTROPY CONTROL OF ECOSYSTEMS

6.1 INTRODUCTION

The derivation of Automatic Control from Global Entropy did not generate any new formulations of the subject. It simply generalizes its meaning and relates it to other global systems that use entropy as a residual energy. Thus Automatic Control may be applied to improve the behavior of societal, economic, environmental and similar systems that affect our modern society (Brooks, Wiley 1988, Coveney, Highfield 1995, Prigogine 1980, 1996, Rifkin 1989).

Recently, Automatic Control has been applied, first, in technological systems. The Germans applied it to the propulsion of the V_2 rockets during the second World War. In the 1950's, Werner von Braun successfully transferred the technology to the space program for the trajectory control of space vehicles. Then new technical applications were developed, including control of communication satellites for the stabilization of their orbits, and their satellite tracking antennas. The technology of Automatic Control, has been recently applied to industrial processes leading to the "factory of the future". This involves the automatization of production, assembly, quality control which can be interpreted as a minimization of the entropy of cost. This has been accomplished by the development of autonomous robots and their use in different stations of product manufacturing.

Automatic Control has been used in developed countries, to limit the industrial pollution of ecological systems. This was done by controlling and processing the waste such that the pollution would stay within acceptable limits. Typical examples are the cleaning of the Hudson river, lake Erie etc. where the entropy of the pollution of the environment is minimized (Prigogine 1980, Rifkin 1989)(see Fig. 5.1).

Automatic Control may find fruitful applications to a country's economy. The flow of currency, the stock market, the marketability of products, the budgetary allocations, the financing of projects and other similar items may be optimally controlled to reduce the danger of financial disasters and the increase of the national debt, which may be interpreted as the economic entropy of the system.

6.2 THE ENVIRONMENTAL SYSTEMS

A very interesting application, recently proposed, is separation and processing for recycling of city waste deposited in dumps around populated areas. Intelligent robots may search and separate biodegradable materials, plastics, paper, and metallic objects for

Fig. 6.1 The Ideal Ecosystem

recycling or destruction. The pollution of the ecosystem, expressed by its entropy may be considerably reduced by this use of Automatic Control (Saridis 1995).

The biological purification of the drinking water is another possible application of Automatic Control in order to reduce the entropy generated by parasites in the source of the water

The length of the list of possible applications of the theory of Automatic Control using the entropy formulation may be further extended by considering problems of our modern society. The proposed solutions express our desire of reducing the maximum entropy of the system which is generated when we implement improvements of the quality of life. Models of those systems are usually found in their literature, which sometimes is difficult to interpret analytically, as required by the theory of Automatic Control (Saridis 1995). This is going to be attempted in the future sections of this paper.

6.3. ANALYTIC FORMULATIONS

There exists a huge literature with analytic formulations of the problems that concern our modern societies, like ecology, environment, biochemical systems, econometrics, and other applications. A good summary of those systems can be found in Singh's **Systems and Control Encyclopedia** (Singh 1987) which served as a source for the development of the material associated with the application of Optimal Control for the reduction of the Global Entropy generated by the work produced by humans in order to improve the quality of life.

The analytic models considered here are for;

- Ecosystems
- Biochemical Systems
- Environmental and Pollution
- Econometric Systems

In most of these cases, the models have been modified to introduce the human effort as a random control variable to be optimized. These models may not be the most general or the most popular ones, since controversies exist among the various researchers, however they are representative to demonstrate the idea of improving the quality of our world by reducing the entropy generated by the work produced.

6.3 ECOLOGICAL SYSTEMS

A bilinear model proposed by Brewer (Singh 1987), is used as a model of population dynamics

$$dx/dt = [-\lambda + v(t)]x(t) + \lambda/P_0 + Bu(t) \tag{6.1}$$

or in a general form

$$dx/dt = f(x,v) + Bu(t) \tag{6.2}$$

where x = 1/P, P the number of individuals in the population, v(t) the birth rate minus the death rate, and u(t) the influence of human intervention. λ, P_0, are appropriate constant vectors and matrices and B is the human intervention gain matrix with elements 0 or 1 depending on the participation of that particular component.

Intercompartmental models arise with the transport of materials between compartments or competition of species;

$$dx/dt = Ax(t) + v(t)Nx(t) + Bu(t) = f(x,v) + Bu(t) \tag{6.3}$$

where $x(t) = [x_1,...x_n]$ are compartmental concentrations u(t) is the active transport, and u(t) the influence of human intervention. A, N, and B are appropriate constant matrices. In the case of n=2, the problem may be viewed as the predator-prey game, and equation (6.2) is reduced to the Lotka-Volterra model [Siljiak, Leslie in (Singh 1987)];

$$dx/dt = \begin{bmatrix} ax_1 - bx_1^2 - cx_1x_2 \\ gx_2 - ex_2^2/x_1 \end{bmatrix} + Bu(t) = f(x_1,x_2) + Bu(t) \tag{6.4}$$

and a,b,c,e, and g are constants.

6.4 BIOCHEMICAL SYSTEMS

One class of biochemical systems deals with metabolic or genetic oscillators, dealing with enzyme activities [Rapp in (Singh 1987)]. The form of their equations is;

$$dx/dt = \begin{bmatrix} g(x) - b_1x_1 \\ x_{j-1} - b_jx_j \end{bmatrix} + Bu(t) = f(x) + Bu(t) \qquad j = 2,..n \tag{6.5}$$

where $f(x) = (1+x^p)/(k+x^p)$ and x is the enzyme induction and u(t) the influence of human contribution in waste. The solution of such equations produce different types of catastrophes that explain the jump phenomena of the equilibrium of the system. Other similar forms of equations, available in the literature, describe other types biochemical processes.

6.6 ENVIRONMENTAL AND POLLUTION MODELS

Environmental analytic models aim to the description of concentration of pollutants (species) in the air or the earths' water resources. If $x=\{x_i\}$ represents the concentration of the pollutants, then the partial differential equation describing this concentration is;

$$\partial x(Try)/\partial t = £_y x(Try) + R(x(Try)) + S(y,t) + Bu(t) \tag{6.6}$$

where y is the spacial variable, R(x(try)) is the rate of generation of species, S(y,t) is the rate of addition of species and u(t) the influence of human intervention. $£_y$ is the

differential operator with respect to y. In the special case of SO_2 pollutant only, the equation is simplified by

$$\partial x'/\partial t = a\partial^2 x'/\partial y^2 + w(y,t) + Bu(t) \tag{6.7}$$

where $x'=x-\langle x\rangle$ with $\langle x\rangle$ the monthly mean concentration satisfying a similar to (6.7) equation. The observation equation for the I sensor, is given by;

$$z(y^i,t) = x'(y^i,t) + v(y^i,t) \tag{6.8}$$

Since no direct measurement methods for x are available, optimal estimates of x are produced using Kalman filter methods [Sawaragi, Omatu in (Singh 1987)];

$$\partial \hat{x}(try/\tau)/\partial t = £_y \hat{x}(try/\tau) + S(y,t) + Bu(t) = f(\hat{x},y,t) + Bu(t) \tag{6.9}$$

The method has given pretty good results for one hour ahead prediction from 4 sensor location with total error squared of 10.63 (Singh 1987).

6.7 ECONOMETRIC MODELS

Econometric analytic systems have been in the literature for a long time, with the most famous one the Leontief model. These are abstract models that deal mostly with equilibrium situations of the economy (Singh 1987). The dynamic models that generate them are of the generic type of:

$$dx/dt = f(x(t),t,v(t),\theta) + Bu(t) \tag{6.10}$$

where x(t) is an vector economic situation, t is time, v(t) is a random error, θ is a parameter vector, and u(t) the influence of human intervention. Such models fit nicely our formulation of models and are designed to predict various dynamic economic situations.

6.8 THE OPTIMAL CONTROL FOR GLOBAL ENTROPY

In the deliberation of the four characteristic systems selected to represent the current problems of human intervention in the welfare of our globe, the following general model was selected to represent analytically all the above systems:

$$dx/dt = f(x(t),t,v(t)) + Bu(t) \tag{6.11}$$

where x(t) is the state variable, v(t) is noise of the system, u(t) is the human intervention and $f(\cdot,\cdot,\cdot)$, B are the function of the system and a constant matrix respectively. This general model simplifies the derivation of the global entropy for all systems, where the appropriate function should replace f in each case.

The work produced by the human intervention generates the energy expressed by:

$$V(u) = E\{J\} = 1/2\ E\{\int_0^T \|u\|_R^2\, dt\} \tag{6.12}$$

Where R is a weighting matrix and T is the duration depending on the process. One now may follow the same procedure used in Chapter 2 of this document in order to show that the maximum entropy, according to the second axiom of Thermodynamics, proportional to the energy of human intervention, can be optimized by the use of Optimal control. Let H be the Entropy of the models, with unspecified u(t) and p(u(t)) its probability density:

$$H(x_0,u(x,t),p(u)) = H(u) = -\int_{\Omega x0}\int_{\Omega x} p(x_0,u)\ln p(x_0,u)\, du dx_0 \tag{6.13}$$

Then applying the second Law of Thermodynamics, one finds p(u) that maximizes H:

$$\partial H/\partial p = 0,; \quad \partial^2 H/\partial p^2 < 0; \tag{6.14}$$

which yields the worst case probability density and maximum entropy respectively;

$$p(u) = e^{-\lambda-\mu V(u(x,t),x0,t0)}$$

$$H(u) = \lambda + \mu E\{V(u(x,t),x_0,t_0)\} \tag{6.15}$$

The corresponding minimum value with respect to u(x,t), represents the optimal human intervention to the appropriate representative system. As in Chapter 2, this condition implies Theorem 2.1,

$$dH/du = \partial H/\partial u = 0, \qquad dH^2/du^2 > 0$$

$$\text{Min}[\ H(u^*)] \approx \text{Min}_u[\{V(u(x,t),x_0,t_0)\}] \tag{6.16}$$

As in Chapter 2, the Hamilton-Jacobi-Bellman equation may be obtained from the condition of Incompressibility in time of Probability Density:

$$dp/dt = 0 \tag{6.17}$$

which for the selected p(u(x,t)) and for u*, V(u*)=Min_u {V(u(x,t),x_0,t_0)). yields

$$\partial V/\partial t + \underset{u}{\mathrm{Min}}\ [\partial V^T/\partial x\cdot f(x,u,t) + L(x,u,t) + \tfrac{1}{2}\mathcal{L}\{Tr(\partial V^2/\partial x^2 Qgg^T)\}] = 0;$$
$$V(x(T),T) = 0. \tag{6.18}$$

Proceeding with the minimization and solving for an appropriate V for each system

$$u^*(x,t) = -\ R^{-1}[\partial V^T/\partial x]\ B$$

$$\partial V/\partial t + E[\partial V^T/\partial x\cdot f(x,v,t) - \tfrac{1}{2}B^T[\partial V/\partial x]\ R^{-1}[\partial V^T/\partial x]\ B + \tfrac{1}{2}\pounds[Tr[\partial V^2/\partial x^2 Qgg^T]] = 0 \tag{6.19}$$

This methodology applies to more than the four systems, used as sample cases, and simply demonstrates a philosophy that Optimal Control can be used in an effective way of reducing the waste generated by human work represented by the irreversible entropy of the world.

6.9 CONCLUSIONS

It is obvious that our view of the laws of nature have been changed. Great scientists, like the Nobel price winner I. Prigogine (1980) have pointed out the uncertainty governing the models of natural phenomena, and the need of a probabilistic description of our observations of nature. What has been Heisenberg's Principle of Uncertainty has developed into a physical reality of all natural phenomena, leaving the deterministic models of Newtonian mechanics as practical approximations for the every day life. This is the view taken in the present paper, where Control theory has been reformulated under the uncertainty of the design and Entropy has been used as a measure of quality. This permits us to bridge the gap between physics and life-sciences, and use powerful existing powerful analytic methods to manage and control societal and environmental problems. The next step in the development of this approach is to collect experimental data from such processes to identify the parameters associated with the proposed models and apply them to the proposed algorithms to verify their validity.

A major contribution of this paper is the introduction of the **human interaction term** u(t) in the equations of physical systems. Thus the theory of Entropy minimization which was derived from optimal control theory can be applied to minimize the effects of human pollution to the above environments.

6.10 REFERENCES

Brogan W. L., (1974) ***Modern Control Theory*** Quantum Publishers New York N.Y.

Brooks D. R., Wiley E. O., (1988) ***Evolution as Entropy*** University of Chicago Press, Chicago Il.

Coveney P., Highfield R. (1995), ***Frontiers of Complexity*** Ballantine Books, New York NY.

Prigogine Ilya, (1980), ***From Being to Becoming*** , Freeman and Company, San Francisco.

Prigogine Ilya, (1996), ***La Fin des Certitudes***, Editions Odile Jacob, Paris France

Rifkin J., (1989) ***Entropy into the Greenhouse World*** Bantam Books New York.

Saridis G. N, (1995) ***Stochastic Processeses, Estimation, and Control: the Entropy Approach***, John Wiley and Sons, New York.

Singh Madan G. (editor), (1987) ***Systems and Control Encyclopedia; Theory, Technology,Applications***, Vol. 1-8, Pergamon Press, Oxford UK.

CHAPTER 7

A CASE STUDY ON OPTIMAL CONTROL OF INTELLIGENT SPACE TRUSS ASSEMBLY

7.1 INTRODUCTION

The theory of Intelligent Control systems lends an excellent case study that uses Entropy to define a uniform measure of performance of the three levels of an Intelligent Machine. It was developed for optimal control of a space truss assembly at the Center of Intelligent Control Systems for Space Exploration (CIRSSE) at the Rensselaer Polytechnic Institute in 1987-1992.. The importance of such an application and its unique structure seems most suitable for such a construction. The part of the machine that was implemented and successfully tested were the two lower levels e.g., the Coordination and the Execution levels as a first phase of an ambitious project: the organization level was assumed on-line by a human operator in this phase and shown in Fig. 7.1.

The theory of Intelligent Control systems, which was developed by Saridis for this occasion, combines the powerful high-level decision making of the digital computer with advanced mathematical modeling and synthesis techniques of system theory. The original version was using linguistic methods of dealing with imprecise or incomplete information, and was reported by Saridis and Valavanis (1988). Further developments utilized neural and Petri nets, with entropy as measures of the cost of execution (Saridis 1996).

This produces a unified approach for the design of **Intelligent Machines**. As a reminder, the theory may be thought of as the result of the intersection of the three major disciplines of Artificial Intelligence, Operations Research, and Control Theory (see Fig. 3.1). This application is aimed at establishing Intelligent Controls as an engineering discipline, with the purpose of designing Intelligent Autonomous Systems of the future.

The control intelligence, is again hierarchically distributed according to the **Principle of Precision with Decreasing Intelligence (IDI)**, evident in all hierarchical management systems (Saridis 1988). The analytic functions of an Intelligent Machine are implemented by Intelligent Controls, using Entropy as a measure. The resulting structure is composed of three basic levels of controls, each level of which may contain more than one layer of tree-structured functions, as in Chapter 3.

The organization level; is modeled after a Boltzmann machine for abstract reasoning, task planning and decision making;

The coordination level; It is composed of a number of Petri Net Transducers supervised, for command exchange, by a dispatcher, which also serves as an interface to the

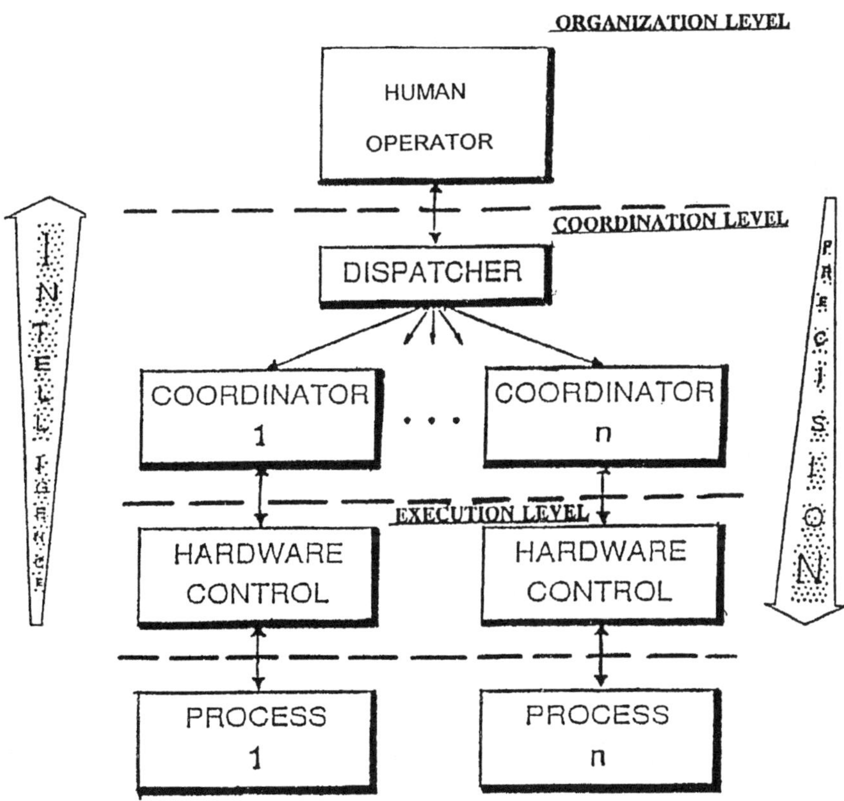

Fig. 7.1 The Structure of the Truss Construction Intelligent Machine

Organization level;

The execution level; includes the sensory, planning for navigation and control hardware which interacts one-to-one with the appropriate Coordinators, while a VME bus provides a channel for database exchange among the several devices.

The functions involved in the upper levels of an intelligent machine are imitating functions of human behavior and may be treated as elements of knowledge-based systems. Actually, the activities of planning, decision making, learning, data storage and retrieval, task coordination, etc., may be thought of as knowledge handling and management (Valavanis Saridis 1992).

Such a system was implemented to perform autonomously construction of NASA's Space Station Freedom at the CIRSSE Laboratories at RPI. The Entropy measures of performance for the optimization of the last two levels of the system are shown in Figure 7.2.

7.2 THE ARCHITECTURE OF THE ORGANIZATION LEVEL

As mentioned earlier the organization level of this structure is replaced by a human operator. Therefore there is no analytic model for this level.

7.3 THE ARCHITECTURE OF THE COORDINATION LEVEL

The *Coordination level* is a tree structure of *Petri Net Transducers* as coordinators, proposed by Wang and Saridis (1990) with the Dispatcher as the root and is shown in Figure 7.2. The commands generated by the Human Organizer defining every specific plan is transmitted, asynchronously, to the Dispatcher of the Coordination level, modeled by a reconfigurable Petri Net Transducer (PNT). The function of the Dispatcher is to interpret the plan and assign individual tasks to the other coordinators, monitor their operation, and transmit messages and commands from one coordinator to another as needed. As an example, a command is sent to the vision and sensing coordinator to generate a model of the environment, the coordinates of the objects for manipulation to be tabulated, and then transmitted to the motion coordinator for navigation and motion control. This command is executed by having each transition of the associated Petri Nets to initialize a package corresponding to a specific action. These packages are stored in short memories associated with each of the coordinators.

The rest of the coordinators have a fixed structure with alternate menus available at request. They communicate commands and messages with each other, through the Dispatcher. They also provide information about reception of a message, data memory location, and job completion.

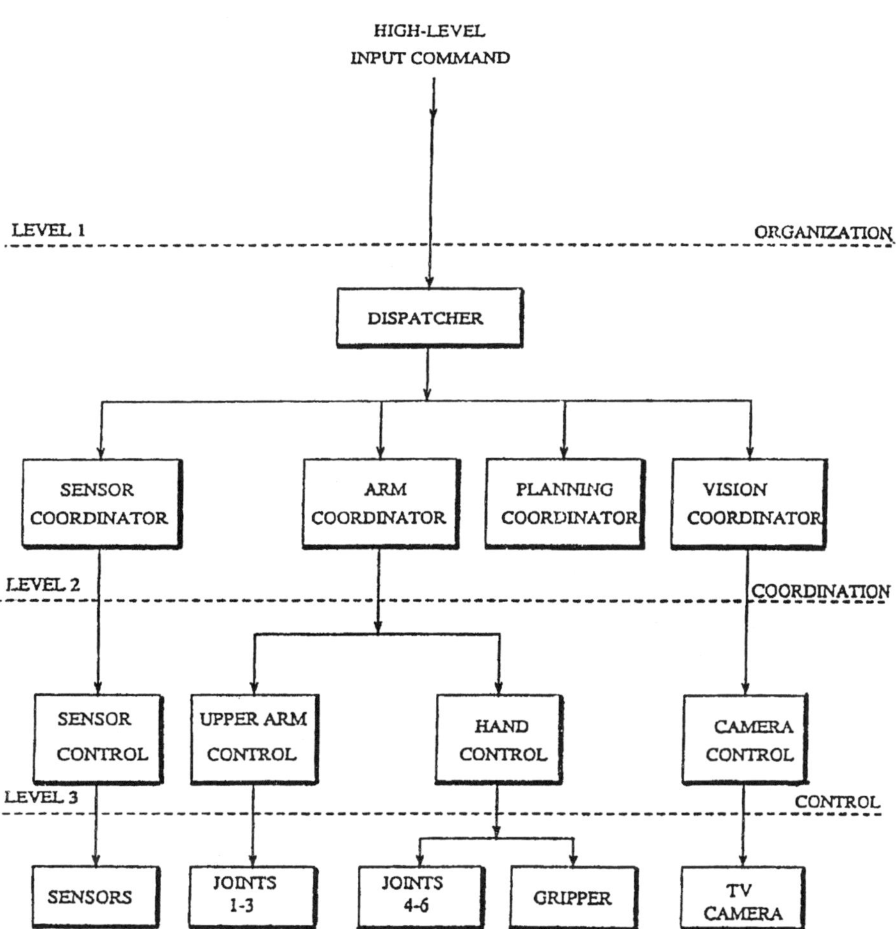

Fig. 7.2 Block Diagram of the Truss Construction

No data is communicated at the Coordination level, since the task planning and monitoring may be located in a remote station, and such an exchange may cause a channel congestion. A preferred configuration (see Fig. 3.9) for such situations is that the coordinators with a local dispatcher may be located with the hardware at the work site, while a remote dispatcher, connected to the organizer, interacts with local one from a remote position. This concept simplifies the communication problem considerably, since only short messages are transmitted back and forth through a major channel between local and remote stations, requiring a narrow bandwidth. An example of the effectiveness of such an architecture may be demonstrated in space construction, where robots work in space while task planning and monitoring is done on earth.

Even though, there is no limitation to the number of coordinators attached to the Dispatcher, only the following ones are planned for an Intelligent Robot for space truss assembly.

Vision Coordinator. This device coordinates all the sensory activities of the robot, with cameras and lazers, and generates information of the world model in Cartesian coordinates.

Sensory Coordinator. This device coordinates mainly the activities of proximity sensors, for fine motion control.

Motion Control Coordinator. This device receives control, object and obstacle information and uses it to navigate and move multiple robotic arms and other devices, for object manipulation and task execution. It also assigns the appropriate operations on the data acquired for the desired application.

Entropy measures, expressing the system reliability, are developed by McInroy and Saridis (1991), at each coordinator such that they may be used to minimize the complexity and improve the reliability of the system.

7.4 THE ANALYTIC MODEL

Petri nets have been proposed in Chapter 3, as possible devices to communicate and control complex heterogenous processes. These nets provide a communication protocol among stations of the process as well as the control sequence for each one of them. Abstract task plans, suitable for many environments are generated at the organization level by a grammar created by Wang and Saridis (1990):

$$G = (N, \Sigma_o, P, S)$$

where

$N = \{S, M, Q, H\}$ = Non-terminal symbols
$\sum_o = \{A_1, A_2, ... A_n\}$ = Terminal Symbols (activities)
P = Production rules

Petri Net Transducers (PNT) proposed first by Wang and Saridis, are Petri net realizations of the *Linguistic Decision Schemata* introduced by Saridis and Graham (1984), as linguistic decision making and sequencing devices. They are defined as 6-tuples:

$$M = (N, \sum, \delta, G, \mu, F)$$

where

$N = (P, T, I, O)$ = A Petri net with initial marking μ,
$\sum$ = a finite input alphabet
δ = a finite output alphabet
σ = a translation mapping from $T \times (\sum \cup \{\lambda\})$ to finite sets of δ^* and $F \subset R(\mu)$ a set of final markings.

A *Petri Net Transducer (PNT)* is a Petri Net with input and output tapes. It communicates with the other levels through input and output languages called *Petri Net Languages (PNL)*. In addition to its on-line decision making capability PNT's have the potential of generating communication protocols, learning by feedback, ideal for the communication and control of coordinators and their dispatcher in real time. Their architecture for the particular Truss Construction in space is given in Figure 7.3, and may follow a scenario suitable for the implementation of an autonomous intelligent robot.

Decision making in the coordination structure is accomplished by *Task Scheduling* and *Task Translation* (Wang and Saridis 1990).

The sequence of events transmitted from the organization level is received by the dispatcher which requests a world model with coordinates from a vision coordinator. The vision coordinator generates appropriate database and upon the dispatcher's command communicates it to the planning coordinator which set a path for the arm manipulator. A new command from the dispatcher sends path information to the motion controller in terms of end points, constraint surface and performance criteria. The vision coordinator is then switched to a monitoring mode for navigation control, and so on.

The PNT can be evaluated in *real-time* by testing the computational complexity and system reliability of their operation which may be expressed uniformly in terms of entropy.

Feedback information is communicated to the coordination level from the execution level

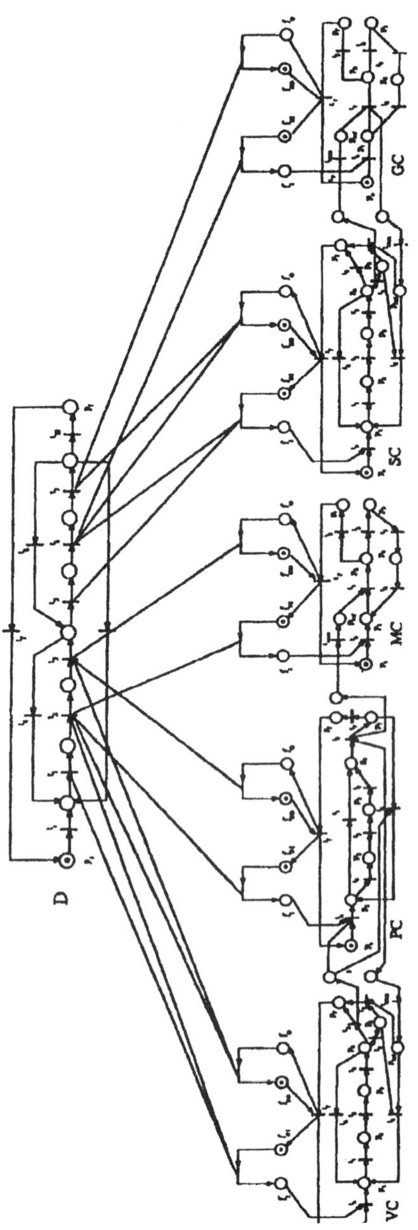

Fig. 7.3 Petri Net Diagram of the Coordination Level for Truss Cnstruction

during the execution of the applied command. Each coordinator, when accessed, issues a number of commands to its associated execution devices (at the execution level). Upon completion of the issued commands feedback information is received by the coordinator and is stored in the *short-term memory* of the coordination level.

This information is used by other coordinators if necessary, and also to calculate the individual, accrued and overall accrued costs related to the coordination level. Therefore, the feedback information from the execution to the coordination level will be called *on-line, real-time* feedback information, and has learning properties (Lima and Saridis 1996). The performance estimate and the associated subjective probabilities are updated after the k_{ij}-th execution of a task $[(u_t,x_t)_i,S_j]$ and the measurement of the estimate of the observed cost J_{ij}:

$$J_{ij}(k_{ij}+1) = J_{ij}(k_{ij})+\beta(k_{ij}+1)[J_{obs}(k_{ij}+1)-J_{ij}(k_{ij})] \qquad (7.1)$$
$$P_{ij}(k_{ij}+1) = P_{ij}(k_{ij})+\mu(k_{ij}+1)[\Gamma_{ij}(k_{ij}+1)-P_{ij}(k_{ij})]$$

where

$$\Gamma_{ij} = \begin{cases} 1 & \text{if } J_{ij} = \text{Min} \\ 0 & \text{elsewhere} \end{cases}$$

and β and μ are harmonic sequences. Convergence of this algorithm is proven in Saridis and Graham (1984).

The *learning process* is measured by the entropy associated to the subjective probabilities. If

$$H(C) = H(E) + H(T/E) \qquad (7.2)$$

where H(E) is the environmental uncertainty and H(T/E) is the pure translation uncertainty. Only the last term can be reduced by learning.

7.5 THE ARCHITECTURE OF THE EXECUTION LEVEL.

The *Execution level* contains all the hardware required by the Intelligent Machine to execute a task. There is a one-to-one correspondence between hardware groups and coordinators. Therefore their structure is usually fixed. This level also contains all the drivers, VME buses, short memory units, processors, actuators and special purpose devices needed for the execution of a task. After the successful completion of a job feedback information is gene-rated at this level for evaluation and parameter updating of the whole machine. Complexity dominates the performance of this level. Since *precision* is proportional to *complexity*, it also defines the amount of effort required to execute a task.

It has been shown that all the activities of this level can be measured by entropy, which may serve as a measure of complexity as well. Minimization of local complexity through feedback, may serve as local design procedure.

The localization of data exchange at this level provides a means of efficient remote control of the Intelligent Machine.

The diversity of the hardware in a general purpose Intelligent Machine is too abstract for the Truss Construction project under consideration. Therefore, this work will focus on the special case of a robot designed for space construction like the CIRSSE transporter. The following hardware groups shown in the Laboratory configuration of Fig. 7.4, are available:

<u>The Vision System.</u> This systems consists of two cameras fixed at the ceiling of the lab., two penlight cameras on the wrist of one PUMA arm, and a lazer range finder. They are all controlled by a Datacube with a versatile menu of various hardwired functions and a VME bus for internal communications. The functions assigned to them, e.g. create a world model in Cartesian space, find the fiducial marks on the object to be manipulated, or track a moving object are supported by software specialized for the hardware of the system. Calibration and control of the hardware is an important part of the system. Since we are dealing with information processing the system's performance can be easily measured with entropy. Actual data for visual servoing can be generated on the VME bus and transmitted through the Dispatcher to the Motion Control system. Internal communication on a VME bus, and direct connection with the VME bus of the Motion Control System has been developed as the VSS system.

<u>The Sensory System.</u> This system consists of two proximity sensors that are used for control of fine motion of the manipulator and the gripper. It functions in a manner very similar to the vision system, but it acts independent of the Motion Control and Vision System busses.

<u>The Motion Control System.</u> This system is a unified structure for cooperative motion and force control for multiple arm manipulation. Since motion affects force but not vice versa, motion control is designed independent of the constraint forces, and force control by treating inertial forces as disturbance. Integral force feedback is used with full dynamics control algorithms. The resulting system, named CTOS, was developed as a multiple-processor, VME-bus based, real time robot control system for the CIRSSE 18-degree-of-freedom transporter. It hierarchically integrates the execution algorithms in planning, interaction, and servo control. It works together with the VXWORKS software and provides most of the transformations, and other kinematics and dynamics tools needed for servoing and manipulation. In earlier work it was shown that the control activities can be measured by entropy (Saridis 1988). Therefore the measure of performance of the Motion Control System is consistent with the rest of the architecture of the Intelligent Machine.

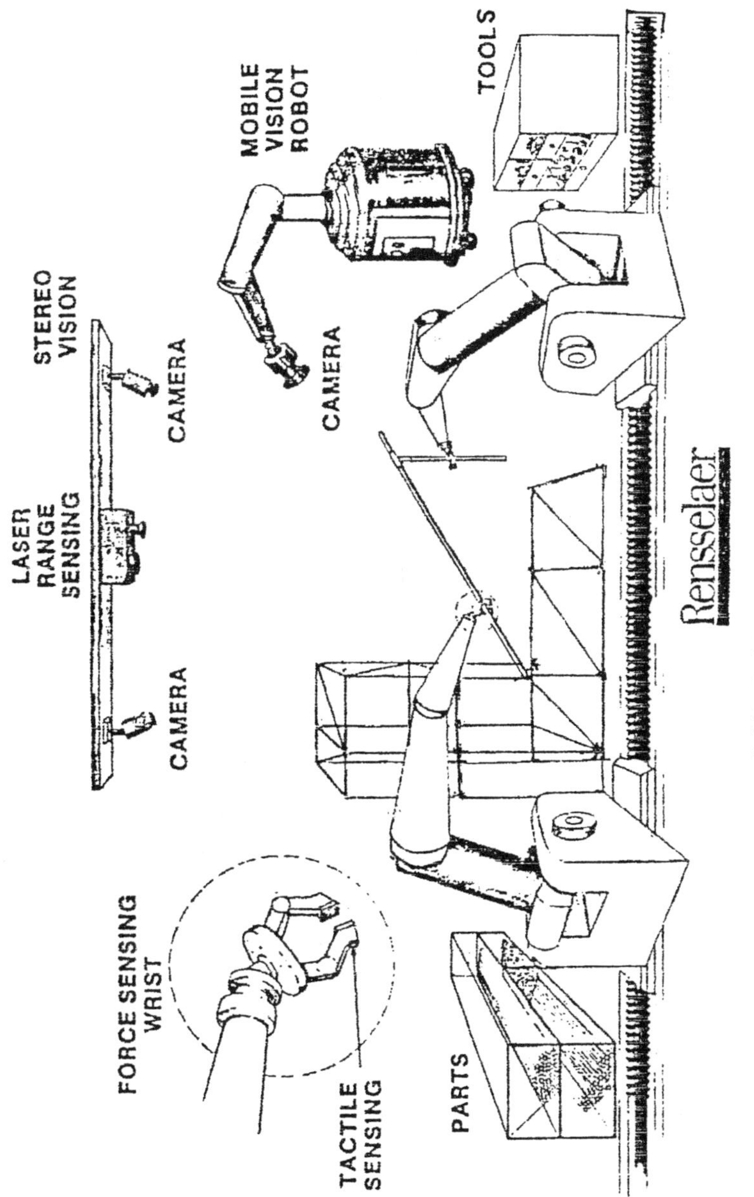

Fig. 7.4 Robotic Assembly of In-Space Structures

7.6 ENTROPY FORMULATION OF MOTION CONTROL.

The cost of control at the hardware level can be expressed as an entropy which measures also the uncertainty of selecting an appropriate control to execute a task. By selecting an optimal control, one minimizes the entropy, e.g., the uncertainty of execution. The entropy may be viewed in this respect as an irreversible lower level energy in the original sense of Boltzmann (1872). The appropriate task for the Truss Assembly system is depicted in Fig. 7.5.

Optimal control theory utilizes the non-negative cost functional of the state of the system $x(t) \in \Omega_x$ the state space, and a specific control $u(x,t) \in \Omega_u \times T$; $\Omega_u \subset \Omega_x$ the set of all admissible feedback controls, to define the performance measure for some initial conditions $x_0(t_0)$, representing a generalized energy function, of the form:

$$V(x_0,t_0) = E\{\int_{t0}^{tf} L(x,t;u(x,t))\, dt\} \tag{7.3}$$

where $L(x,t;u(x,t)) > 0$, subject to the differential constraints, dictated by the underlying process

$$\begin{aligned} dx/dt &= f(x,u(x,t),w,t); \quad x(t_0) = x_0 \\ z &= g(x,v,t); \quad x(t_f) \in M_f \end{aligned} \tag{7.4}$$

where x_0, $w(t)$, $v(t)$ are random variables with associated probability densities $p(x_0)$, $p(w(t))$, $p(v(t))$ and M_f a manifold in Ω_x. The trajectories of the system (7.4) are defined for a fixed but arbitrarily selected control $u(x,t)$ from the set of admissible feedback controls Ω_u.

In order to express the control problem in terms of an entropy function, one may assume that the performance measure $V(x_0,t_0,u(x,t))$ is distributed in u according to the probability density $p(u(x,t))$ of the admissible controls $u(x,t) \in \Omega_u$. *The differential entropy* $H(u)$ corresponding to the density is defined as

$$H(u) = -\int_{\Omega u} p(u(x,t)) \ln p(u(x,t))\, dx$$

and represents the uncertainty of selecting a control $u(x,t)$ from all possible admissible feedback controls Ω_u. The optimal performance should correspond to the maximum value of the associated density $p(u(x,t))$. Equivalently, the optimal control $u^*(x,t)$ should minimize the entropy function $H(u)$. This is satisfied if the density function is selected to satisfy *Jaynes' Principle of Maximum Entropy* (Jaynes 1957), e.g.,

$$p(u(x,t)) = \exp\{-\lambda - \mu V(x_0,t_0;u(x,t))\} \tag{7.5}$$

where λ and μ are normalizing constants.

Puma 560 manipulator performing robotic assembly task:

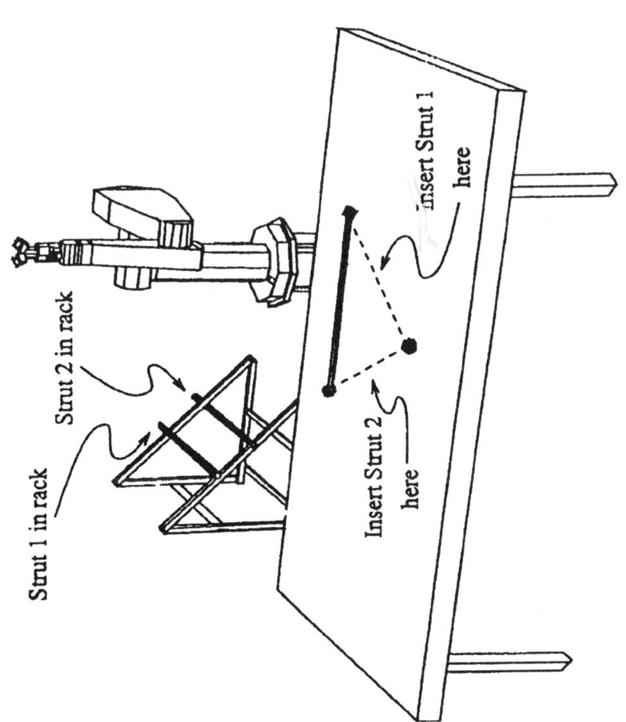

Fig. 7.5 A Case Study

It was shown by Saridis (1988), that the expression H(u) representing the entropy for a particular control action u(x,t) is given by:

$$H(u) = \int_{\Omega u} p(x,t;u(x,t))V(x_0,t_0;u(x,t))\,dx = \lambda + \mu V(x_0,t_0;u(x,t)) \tag{7.6}$$

This implies that the average performance measure of a feedback control problem corresponding to a specifically selected control, is an entropy function. The optimal control $u^*(x,t)$ that minimizes $V(x_0,t_0;u(x,t))$, maximizes $p(x,t;u(x,t))$, and consequently minimizes the entropy H(u).

$$u^*(x,t) : E\{V(x_0,t_0;u^*(x,t))\} = \min \int_{\Omega u} V(x_0,t_0;u(x,t))p(u(x,t))dx \tag{7.7}$$

This statement is the generalization of a theorem proven by Saridis (1988), and establishes equivalent measures between information theoretic and optimal control problem and provides the information and feedback control theories with a common measure of performance.

7.7 ENTROPY MEASURE OF THE VISION SYSTEM.

The Intelligent Control theory, designed mainly for motion control, can be implemented for vision control, path planning and other sensory system pertinent to an Intelligent Machine by slightly modifying the system equations and entropy functions as in Chapter 3. After all one is dealing with real-time dynamic systems which may be modeled by a dynamic set of equations.

A *Stereo Vision* system of a pair of cameras mounted at the end of a robot arm, may be positioned at l=1,..N different view points to reduce problems with noise, considered one at a time due to time limitations. The accuracy of measuring the object's position depends upon its relative position in the camera frame. Consequently, each viewpoint will have different measurement error and time statistics. These statistics may be generated to define the uncertainty of the measurement of the Vision system as in McInroy and Saridis (1991).

For a point c of the object, the measurement error of its 3-D position in the camera coordinate frame e_{pc} is given by:

$$e_{pc} = M_c\, n_c \tag{7.8}$$

where n_c is the 3-D image position errors, and M_c an appropriate 3×3 matrix, depending on the position of the object.

The linearized orientation error is given by:

$$\delta = (M^TM)^{-1}M^TM'Fn \tag{7.9}$$

where

δ is the orientation error in the camera frame,
M is a matrix formed from camera coordinate frame positions,
M' is a constant matrix,
F is the matrix formed from the camera parameters and measured positions,
n is the vector of the image position errors at the four points.

A vector containing the position and orientation errors due to image noise is given by:

$$e_c = [e^T_{pc}\delta^T]^T = Ln \tag{7.10}$$

where L depends on the camera parameters and the four measured camera frame positions of the points. The statistics of the image noise n, due to individual pixel errors are assumed to be uniformly distributed. Assuming that feature matching centroids is used by the vision system, its distributions tend to be independent Gaussian, due to the Central Limit Theorem.

$$n \approx N(0,C_v) \text{ and } e_c \approx N(0,LC_vL^T) \tag{7.11}$$

The time which each vision algorithm consumes is also random due to the matching period. Therefore the total vision time, for the ith Algorithm that includes camera positioning time, image processing time, and transformation to the base frame, is assumed Gaussian:

$$t_{vi} \approx N(\mu_{tvi},\sigma^2_{tvi}). \tag{7.12}$$

Once the probability density functions are obtained, the resulting Entropies $H(t_{vi})$, and $H(e_c)$, are immediately obtained for the ith Algorithm (McInroy and Saridis 1991):

$$\begin{aligned} H(t_{vi}) &= \ln\sqrt{2\pi e\sigma^2_{tvi}}) \\ H(e_c) &= \ln\sqrt{(2\pi e)^6\det[C_v]} + E\{\ln[\det L_i]\} \end{aligned} \tag{7.13}$$

The total Entropy, may be used as a measure of uncertainty of the Vision system (imprecision), and can be minimized with respect to the available system parameters:

$$H(V) = H(t_{vi}) + H(e_c). \tag{7.14}$$

7.8 ENTROPY MEASURE FOR THE SENSORY SYSTEM

The proximity sensory system has similar properties as the vision system. Therefore its entropy measure should be similar to eq. (7.14):

$$H(S) = H(t_{vi}) + H(e_c). \tag{7.15}$$

7.9 TOTAL ENTROPY OF THE SYSTEM

Since the Coordination level works on a different time scale than the Execution level, the total Entropy of the system is obviously the sum of the partial Entropies of the Coordination H(C), the Motion Control H(MC), the Vision system H(V), and the Sensory system H(S).

$$H(TOTAL) = H© + H(MC) + H(V) + H(S) \tag{7.16}$$

By selecting individual controls to minimize the total Entropy H(TOTAL) one obtains the optimal truss assembly procedure for space construction Fig.7.6, is a photograph of the experimental system for Truss Assembly in space designed and operated at the Center for Intelligent Robotic Systems for Space Exploration(CIRSSE) at the Rensselaer Polytechnic Institute in 1993.

7.10 CONCLUSIONS

The Truss Assembly in space project, developed at CIRSSE of RPI, was used as a case study, of the application of Intelligent Control. It demonstrated the usefulness of using Entropy as a unifying measure, for optimal control of a process, combining various disjoint disciplines. This presents a complete justification of the theoretical work presented in this book, and other publications

More importantly, it has established that the cost of performance, expressed as entropy, is an irreversible phenomenon in time, since such a construction assumes energy consumption in space flights, visual recognition, decision making and control activities that cannot be reversed in time. It also stands as a link with other contemporary attempts to introduce entropy as the unifying measure in disciplines like biotechnology, ecology, pollution control, economics and manufacturing.

Finally, as a caveat, it shows the way to the future of the human race to face the problem of overpopulation and over pollution of the earth by colonizing other planets in space. It demonstrates the capabilities of Intelligent Machines to assist humans in this adventure.

7.11 REFERENCES

Boltzmann, L. (1872) "Further Studies on Thermal Equilibrium between Gas Molecules" *Wien Ber.*, **66**, p. 275.

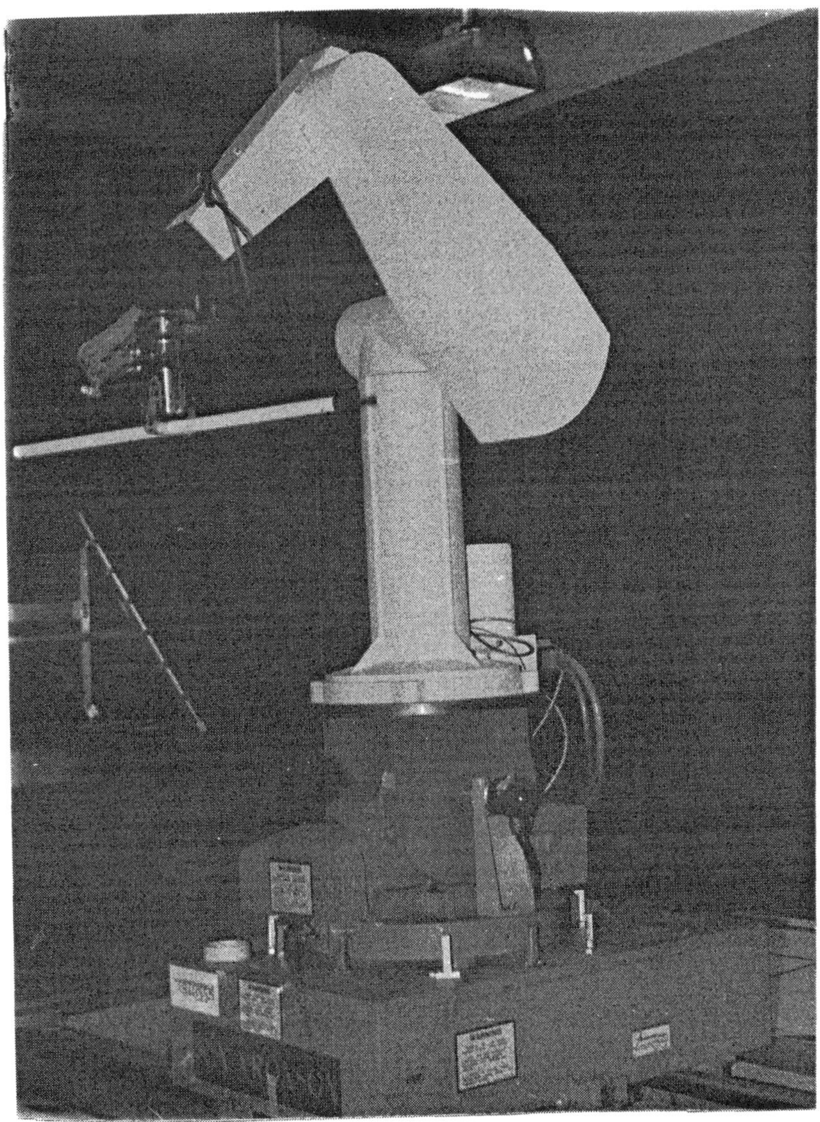

Fig.7.6 Truss Construction in the CIRSSE Laboratory, RPI

Jaynes, E.T. (1957), "Information Theory and Statistical Mechanics", *Physical Review*, Vol.4, pp. 106.

Lima, P. U., Saridis G.N.,(1996), ***Design of Intelligent Control Systems based on Hierarchical Stochastic Automata.*** World Scientific, Singapore

McInroy J.E., Saridis G.N.,(1991), "Reliability Based Control and Sensing Design for Intelligent Machines", in ***Reliability Analysis*** ed. J.H. Graham, Elsevier North Holland, N.Y.

Saridis, G. N. (1983), "Intelligent Robotic Control", *IEEE Trans. on Automatic Control*, Vol. 28, No. 4, pp. 547-557, April

Saridis, G. N. (1988), "Entropy Formulation for Optimal and Adaptive Control", *IEEE Transactions on Automatic Control*, Vol. 33, No. 8, pp. 713-721, Aug.

Saridis G. N. (1996), "Architectures for Intelligent Controls" in ***Intelligent Control Systems***, Edited by Madan. M. Gutta and Nares K. Sinha, Chapter 6, pp.127-148, IEEE Press.

Saridis, G.N. and Valavanis, K.P. (1988), "Analytical Design of Intelligent Machines", *AUTOMATICA the IFAC Journal, 24*, No. 2, pp. 123-133, March.

Wang, F., Saridis, G.N. (1990) "A Coordination Theory for Intelligent Machines" *AUTOMATICA the IFAC Journal, 35*, No. 5, pp. 833-844,Sept.

Valavanis, K.P., Saridis, G.N., (1992), ***Intelligent Robotic System Theory: Design and Applications***, Kluwer Academic Publishers, Boston, MA.

CHAPTER 8.

CONCLUSIONS

8.1 INTRODUCTION

Entropy has been a cornerstone of the theory of Thermodynamics. The general idea, introduced by the Second Law of Thermodynamics, is that in order to produce work, e.g. creating order where there was disorder, requires the consumption of useful energy thus reducing its supply, and increasing the entropy in order to keep the total energy of the system constant, according to the First Law of Thermodynamics (Hawking 1997). Boltzmann (1872), with his statistical theory, introduced entropy as a measure of the cost of consumption of the useful energy of the system. Subsequently, Shannon (1963) used this idea to measure the capacity of a communication channel. Since then the concept of entropy has expanded in many directions in Modern Science and Ecology (Prigogine 1980, 1996, Brooks and Wiley 1988, Rifkin 1989).

Since 1973, Saridis has been working on a Hierarchically Intelligent Control theory, that utilizes entropy as the cost of performance of such a system. This theory grew to include other engineering applications, such as reliability, manufacturing, ecological and pollution systems control, biotechnology etc. Even though the engineering applications may sound simplistic for the sophisticated scientist, it bears a lot of similarities and ties up beautifully with the modern trends of Science. This original work of the last 25 years, is what has been exposed in the present volume, with analytic expressions and examples unifying various approaches in cognitive engineering.

8.2 IRREVERSIBILITY OF PROCESSES

There is a lot of discussion lately of the irreversibility of processes (Prigogine 1980, 1996, Brooks and Wiley 1988) as well as the arrow of time (Coveney and Highfield 1990) on account of entropy. It refers to the one way evolution of processes in time. The concept, essential in life sciences, was extended to control engineering as shown in Chapter 1.The idea of irreversibility as the objective uncertainty of a-priori solution, was introduced in the designer's problem; it was reintroduced when considering that control produces useful work on a system which generates the cost of performance as entropy (Saridis 1985). This irreversibility is interpreted by considering that, when the cost of performance is paid, the system cannot be recovered. This property can be witnessed by visiting the junkyards of old automobiles: new cost must be paid to recover the metal of the old wrecks or in the construction of Space Stations, where the energy of space transportation cannot be recovered.

8.3 THE GLOBALIZATION OF ENTROPY

As mentioned in the Introduction Entropy is becoming very popular in most branches of modern sciences. One of the reasons is that it represents in the modern analytic world, a very compact, single numbered, inclusive measure of the cost of performance, instead of the more complex probability density functions, representing its uncertainties. On the other hand, entropy represents a necessary evil, an energy that is the result of our efforts to improve the quality of human lives, their scientific discoveries, and finally their minds.

Entropy may be thought of as the unifying element that brings together not only modern physics, chemistry, and life-sciences but also engineering sciences that was always considered as a poor relative. The beginning was made with intelligent control engineering and could be extended as well to other branches. The interpretation of natural phenomena through entropy sounds so "natural" that they are automatically accepted. Engineering provides a means of what to do with the arising situations to minimize their destructive effects.

8.4 ENTROPY AND CONTROL ENGINEERING AND CHAOS

The reason for preparing this volume is to summarize the work of the author on entropy on control engineering, and to establish its relation to the sciences and life-sciences. Even though the ideas were generated rather randomly through the years, they appear herein in their natural progression.

The concepts of entropy chaos and control engineering were introduced, to familiarize the reader with the key notions of this book. A new derivation of the theory of optimal estimation and control were presented, by associating the cost of performance with entropy. This new theory produced two surprising results: first the derivation of the Generalized Hamilton-Jacobi-Bellman equation which describes the behavior of control systems for a fixed but arbitrary controller away from the optimal (equilibrium) solution and second the analytic verification of Fel'dbaum's claim of Dual Control that the optimal closed-loop solution is obtained by the separation theorem *minus the active transmission of information.* The theory of Hierarchically Intelligent Control Systems was also reviewed, a topic which was the main thrust of the author's research for many years. Actually this work presented the motivation for using entropy as a measure of the cost function of such disjoint disciplines as control and sensing. The entropy formulation is presented for the three levels of the hierarchy: *Organization, Coordination, and Execution.*

An idea of defining system reliability as entropy was also explored as a first application of this theory. Using tight measures one may come up with simpler expressions which actually separates the cost of reliability from the cost of the experiment. Entropy in the performance of a modern fully automated factory was the next topic. It also demonstrates the application of Hierarchically Intelligent Control to production scheduling. The versatility of Intelligent Control is demonstrated by using methodologies different from the ones described in Chapter 3 to obtain the final goal. Application of entropy derived control to environmental systems is a further application. Here the principle of minimizing the entropy generated by human intervention with the environment is applied demonstrating a way of the waste produced by generating work for the improvement of the quality of living. The systems under consideration are only four, Ecological, Biochemical, Pollution, and Econometric, but the concept can be extended to other non-engineering systems. The application of Hierarchically Intelligent Control to a hardware oriented system; the unmanned construction of trusses for the Space Station was introduced to demonstrate further the connection of the theory to the real world. This project was funded by NASA, as a Center of Excellence and was performed at the Center of Intelligent Control Systems for Space Exploration (CIRSSE) at the Rensselaer Polytechnic Institute between 1987 and 1992. It was unfortunate that the project was terminated due to discontinuation of funding when it was reaching its fruition.

The present work has helped greatly to integrate control engineering problems with the rest of the modern developments in the sciences which modify modern physics and models analytically life-sciences (Prigogine 1996). Entropy has recently introduced the formulation of a control theory away from the optimal solution, a concept different from the presently accepted theories of the Maximum Principle and the Hamilton-Jacobi-Bellman equations, considered as equivalent to an equilibrium solution. This coincides with the views of modern scientists which, with the aid of computers, have discovered the behavior of nonlinear differential equations away from their equilibrium, which has been named Chaos, and explains as well as predicts the behavior of biological systems in real life. The respective results have been obtained in Control Engineering, some time ago, by Saridis and his colleagues (Saridis and Lee 1979, Wang and Saridis 1992, Saridis 1996). In this work, differential equations describing the performance of a system under a fixed but arbitrary controller were obtained away from the equilibrium (optimal) solution using entropy concepts and were analyzed in Chapter 2.

8.5 REMARKS

The entropy approach provided a new methodology of deriving the theory of to control engineering as well as relations to other sciences. The irreversibility is introduced in the sense that the cost of performance is equivalent to entropy. This is compatible with the modern interpretation of life phenomena in life-sciences (Brooks and Wiley 1988), ecology (Rifkin 1989), biochemistry (Prigogine 1980), and justifies Boltzmann's original theory (1872) and his philosophy (Fasol-Boltzmann 1990).

The contribution of this book is the introduction of entropy in Control Engineering and Manufacturing, the development of the Principle of Increasing Precision with Decreasing Intelligence (IDI), and finally the introduction of the human intervention terms in ecological and other systems such that it may be controlled. Finally, real life applications of the theory were introduced to demonstrate its validity. I hope it will be accepted by the engineering and the life-science communities as unifying methodology. Therefore, considerable encouragement is required to continue this work .

8.6 REFERENCES

Boltzmann, L. (1872) "Further Studies on Thermal Equilibrium between Gas Molecules" *Wien Ber.*, **66**, p. 275.

Brooks D. R. and Wiley, E. O. (1988) ***Evolution as Entropy*** University of Chicago Press, Chicago Il.

Coveney, P. and Highfield, R. (1990) ***The Arrow of Time***, Fawcett Colombine New York N.Y.

Fasol-Boltzmann I. M., editor, (1990) ***Ludwig Boltzmann Principien der Naturfilosofi*** Springer Verlag, Berlin Germany.

Hawking, Steven (1997) ***A Brief History of Time*** Bantam Paperbacks, New York N.Y.

Prigogine, Ilya (1980), ***From Being to Becoming*** , Freeman and Company, San Francisco.

Prigogine, Ilya (1996), ***La Fin des Certitudes*** Editions Odile Jacob, Paris France.

Rifkin, Jeremy(1989) ***Entropy into the Greenhouse World*** Bantam Books New York.

Saridis, G. N. (1985), "An Integrated Theory of Intelligent Machines by Expressing the Control Performance as an Entropy", ***Control Theory and Advanced Technology***, ***Vol. 1,*** No. 2, pp. 125-138, Aug.

Saridis, G. N. and Lee, C.S.G. (1979), "Approximation of Optimal Control for Trainable Manipulators", *IEEE* ***Trans. on Systems Man and Cybernetics, Vol.8***, No. 3, pp. 152-159, March

Saridis, G. N. (1988), "Entropy Formulation for Optimal and Adaptive Control", *IEEE* ***Transactions on Automatic Control***, ***Vol. 33***, No. 8, pp. 713-721, Aug.

Saridis, G. N. (1989), "Analytic Formulation of the IDI for Intelligent Machines", ***AUTOMATICA the IFAC Journal***, **25**, No. 3, pp. 461-467.

Saridis, G. N. (1995) ***Stochastic Processeses, Estimation and Control: the Entropy Approach***, John Wiley and Sons, New York.

Shannon C., Weaver W. (1963), ***The Mathematical Theory of Communications***, Aeolian Books , Urbana Ill.

Wang, F. Y. and Saridis G. N., (1992) "Suboptimal Control for Nonlinear Stochastic Systems" ***Proceedings of 1992 Conference on Decision and Control***, Tucson AZ, Dec.

INDEX